自然语言处理实战

主　编　黎远松　赵良军　夏国栋
副主编　黄坤超

西南交通大学出版社
·成都·

内容简介

本书为计算机类专业核心课程"自然语言处理"配套教材。第 1~3 章按照概念的难易程度编排，先介绍自然语言处理的入门知识，再讲述如何使用简单的 Python 程序去分析感兴趣的文本信息。第 4 章是结构化程序设计，用来巩固分散在前面几章中的编程要点。第 5~7 章讲述自然语言处理的标注、分类和信息提取。每一章在两种不同的讲述风格之间切换。一种风格是以自然语言处理为主线，分析语言，探索语言学，在讨论中使用编程的例子，此时会使用尚未系统介绍的 Python 结构，这样读者便可在钻研这些程序如何运行的细节之前了解它们的效能。另一种风格是以 Python 程序设计语言为主线，分析程序，探索算法，而语言学例子扮演配角，同时，部分章节配有阅读材料，用来介绍一些拓展材料和网络资源。

本书可作为普通高等院校计算机科学与技术研究生的教学用书，也可作为从事自然语言处理的科技人员的参考书。

本书得到 2021 年四川轻化工大学产教融合教材出版资助。

图书在版编目（CIP）数据

自然语言处理实战 / 黎远松，赵良军，夏国栋主编.
成都：西南交通大学出版社，2025. 3. -- ISBN 978-7-5774-0164-5

Ⅰ．TP391

中国国家版本馆 CIP 数据核字第 2024H23X65 号

Ziran Yuyan Chuli Shizhan

自然语言处理实战

主　编／黎远松　赵良军　夏国栋	策划编辑／胡　军
	责任编辑／李华宇
	封面设计／GT 工作室

西南交通大学出版社出版发行

（四川省成都市金牛区二环路北一段 111 号西南交通大学创新大厦 21 楼　610031）
营销部电话：028-87600564　　028-87600533
网址：https://www.xnjdcbs.com
印刷：成都蜀通印务有限责任公司

成品尺寸　185 mm×260 mm
印张　19.5　　字数　487 千
版次　2025 年 3 月第 1 版　　印次　2025 年 3 月第 1 次

书号　ISBN 978-7-5774-0164-5
定价　59.00 元

课件咨询电话：028-81435775
图书如有印装质量问题　本社负责退换
版权所有　盗版必究　举报电话：028-87600562

前　言

自然语言，是指人们日常交流使用的语言，如汉语、英语等。相对于编程语言和数学符号这样的人工语言，自然语言随着一代人传给另一代人而不断演化，因而很难用明确的规则来刻画。从广义上讲，自然语言处理（NLP）包含所有用计算机对自然语言进行的操作，从简单的通过计数词出现的频率来比较不同的写作风格，到复杂的完全"理解"人所说的话，至少要能达到对人的话语做出有效反应的程度。

基于 NLP 的技术应用日益广泛。例如，手机和手持式计算机支持输入法联想提示和手写识别；网络搜索引擎能搜到非结构化文本中的信息；机器翻译能把中文文本翻译成英文。通过提供更自然的人机界面和更复杂的存储信息获取手段，语言处理正在这个多语种的信息社会中扮演更核心的角色。

本书实践性强，包含几百个实际可用的例子和分级练习，可作为自然语言处理课程的教材，也可作为人工智能、文本挖掘、语料库语言学课程的补充读物。

本书基于 Python 编程语言及其上的一个名为自然语言工具包（Natural Language Toolkit，NLTK）的开源库，使用时建议先下载 Python 和 NLTK。

本书由四川轻化工大学黎远松、赵良军、夏国栋担任主编，黄坤超担任副主编。具体编写分工如下：第 1、2、3 章由黎远松编写，第 4、5、6 章由赵良军编写，第 7 章由夏国栋编写，阅读材料由黄坤超编写。

由于作者水平所限，书中难免存在疏漏、不妥之处，敬请读者批评指正。

编　者
2024 年 6 月

目 录

第 1 章 基础工具与数据的使用 ·· 1
  1.1 文本和单词 ··· 1
    1.1.1 Python 的安装 ··· 1
    1.1.2 NLTK 的安装 ·· 1
    1.1.3 搜索文本 ··· 2
    1.1.4 词汇计数 ··· 4
  1.2 文本的表示 ··· 6
    1.2.1 列表 ·· 6
    1.2.2 索引列表 ··· 7
    1.2.3 变量 ·· 8
    1.2.4 字符串 ··· 8
  1.3 语言计算 ··· 9
    1.3.1 频率分布 ··· 9
    1.3.2 高频长词 ··· 11
    1.3.3 词语搭配和双连词 ·· 12
    1.3.4 词长计数 ··· 12
  1.4 流程控制 ··· 13
    1.4.1 条件 ·· 13
    1.4.2 列表推导 ··· 15
    1.4.3 控制结构 ··· 15
    1.4.4 条件循环 ··· 16
  小结 ·· 17
  练习 ·· 17
  阅读材料 ·· 26

第 2 章 文本语料库获取 ··· 28
  2.1 文本语料库概述 ·· 28
    2.1.1 gutenberg 语料库 ··· 28
    2.1.2 webtext 和 nps_chat 语料库 ·· 29
    2.1.3 brown 语料库 ·· 30
    2.1.4 多国语言语料库 ··· 32
    2.1.5 文本语料库的结构 ·· 33
  2.2 条件频率分布 ·· 34
    2.2.1 分组条件 ··· 34

2.2.2　按文体计数词汇 34
　　2.2.3　绘制分布图和分布表 35
　　2.2.4　使用双连词生成随机文本 37
2.3　代码重用 38
　　2.3.1　使用文本编辑器创建程序 38
　　2.3.2　函数 39
　　2.3.3　模块 40
2.4　词典资源 41
　　2.4.1　词汇列表语料库 41
　　2.4.2　发音词典 42
　　2.4.3　对照词表 43
小结 43
练习 44
阅读材料 59

第3章　文本处理

3.1　从网络和本地访问文本 67
　　3.1.1　从网络访问文本 67
　　3.1.2　HTML 处理 69
　　3.1.3　本地文件读取 69
　　3.1.4　input 函数 70
　　3.1.5　自然语言处理流程 70
3.2　字符串 71
　　3.2.1　字符串的基本操作 71
　　3.2.2　访问单个字符 73
　　3.2.3　子串访问 74
　　3.2.4　有用的字符串方法 75
　　3.2.5　列表与字符串的差异 75
3.3　使用 Unicode 进行文字处理 76
　　3.3.1　Unicode 简介 76
　　3.3.2　从文件中提取已编码文本 76
3.4　使用正则式检测词组搭配 77
　　3.4.1　使用基本的元字符 77
　　3.4.2　范围与闭包 78
3.5　正则式的应用 80
　　3.5.1　提取字符块 80
　　3.5.2　单词处理 83
　　3.5.3　查找词干 85
　　3.5.4　搜索已分词文本 85

  3.6 规范化文本···87
    3.6.1 词干提取器···87
    3.6.2 词形归并··90
  3.7 用正则式为文本分词···92
    3.7.1 分词的简单方法···92
    3.7.2 NLTK 的正则式分词器···94
  3.8 分割···96
    3.8.1 断句··96
    3.8.2 分词··96
  3.9 数据制表···101
  小结···102
  练习···103
  阅读材料···123

第 4 章 结构化程序设计···129
  4.1 序列···129
    4.1.1 序列的分割··129
    4.1.2 序列的嵌套··130
  4.2 风格···131
    4.2.1 过程风格与声明风格···131
    4.2.2 循环变量的使用··134
  4.3 函数···136
    4.3.1 参数类型检查···136
    4.3.2 函数的设计··138
    4.3.3 作为参数的函数··140
    4.3.4 迭代与递归··141
    4.3.5 filter 和 map 函数··143
    4.3.6 参数的命名··144
  4.4 程序开发···145
    4.4.1 Python 模块的结构··145
    4.4.2 Pdb 调试器···147
  4.5 算法设计···148
    4.5.1 递归与迭代··148
    4.5.2 权衡空间与时间··151
  4.6 Python 库的样例··154
    4.6.1 brown 语料库中特殊情态动词的频率··························154
    4.6.2 WordNet 网络结构的可视化····································156
    4.6.3 CSV 格式的简单词典··157
  小结···158

练习 ··· 158
　　阅读材料 ··· 174

第 5 章　词性标注 ·· 177
5.1　词性标注器 ·· 177
5.2　标注语料库 ·· 178
5.2.1　表示已标注的标识符 ·· 178
5.2.2　已标注语料库的读取 ·· 179
5.2.3　简化的词性标记集 ·· 180
5.2.4　名词 ··· 182
5.2.5　动词 ··· 183
5.2.6　未简化的标记 ·· 185
5.2.7　已标注语料库探索 ·· 186
5.3　使用 Python 字典映射词及其属性 ······································ 190
5.3.1　索引列表和字典 ·· 190
5.3.2　Python 字典 ··· 190
5.3.3　定义字典 ··· 193
5.3.4　递增地更新字典 ·· 195
5.3.5　复杂的键和值 ·· 196
5.3.6　颠倒字典 ··· 197
5.4　自动标注 ··· 199
5.4.1　默认标注器 ··· 199
5.4.2　正则表达式标注器 ·· 201
5.4.3　查询标注器 ··· 202
5.5　N-gram 标注 ··· 204
5.5.1　一元标注(Unigram Tagging) ···································· 204
5.5.2　分离训练和测试数据 ·· 205
5.5.3　N-gram 标注 ·· 206
5.5.4　组合标注器 ··· 208
5.5.5　存储标注器 ··· 209
5.5.6　性能限制 ··· 210
5.5.7　跨句子边界标注 ·· 211
5.6　基于转换的标注 Brill ·· 212
5.7　确定一个词的分类 ··· 214
5.7.1　形态学线索 ··· 214
5.7.2　句法线索 ··· 214
5.7.3　语义线索 ··· 215
　　小结 ··· 215
　　练习 ··· 215

第6章 文本分类 ··· 231

6.1 有监督分类 ·· 231
6.1.1 性别分类 ·· 231
6.1.2 选择正确的特征 ·· 234
6.1.3 文档分类 ·· 237
6.1.4 词性标注 ·· 238
6.1.5 探索上下文语境 ·· 239
6.1.6 序列分类 ·· 240

6.2 有监督分类的典型应用 ··· 243
6.2.1 句子分割 ·· 243
6.2.2 识别对话行为类型 ··· 245
6.2.3 识别文字蕴含 ··· 246

6.3 评估 ··· 247
6.3.1 测试集 ··· 247
6.3.2 准确度 ··· 248
6.3.3 精确度和召回率 ·· 249
6.3.4 混淆矩阵 ·· 249

6.4 决策树 ·· 252
6.4.1 决策树的构造 ··· 252
6.4.2 熵和信息增益 ··· 252

6.5 朴素贝叶斯分类器 ·· 254
6.5.1 核心思想 ·· 254
6.5.2 概率基础 ·· 254
6.5.3 朴素贝叶斯分类器的应用 ·· 255

6.6 最大熵原理 ·· 256
6.6.1 最大熵模型 ·· 257
6.6.2 熵的最大化 ·· 258
6.6.3 生成式分类器对比条件式分类器 ······························· 259

小结 ·· 259
练习 ·· 260

第7章 信息提取 ··· 264

7.1 信息提取概述 ·· 264
7.2 分块 ··· 267
7.2.1 名词短语分块 ··· 267
7.2.2 标记模式 ·· 269
7.2.3 用正则式分块 ··· 269
7.2.4 文本语料库探索 ·· 271

		7.2.5 加缝隙 ·· 272
		7.2.6 块的表示 ·· 273
	7.3 开发和评估分块 ·· 275
		7.3.1 读取IOB格式与CoNLL2000分块语料库 ············· 275
		7.3.2 简单评估和基准 ·· 276
		7.3.3 训练基于分类器的分块器 ···························· 278
	7.4 语言结构中的递归 ·· 281
		7.4.1 用级联分块器构建嵌套结构 ························· 281
		7.4.2 树 ··· 283
		7.4.3 树的遍历 ·· 288
	7.5 命名实体识别 ·· 289
	7.6 关系抽取 ·· 294
	小结 ·· 296
	练习 ·· 297

参考文献 ·· 302

第 1 章　基础工具与数据的使用

1.1　文本和单词

1.1.1　Python 的安装

Python 的安装方法有很多种，本书推荐使用 Anaconda 发行版。该版本集成了许多必要的库，用户可以一次性完成安装。

完成 Python 的安装后，需要先确认 Python 的版本。打开终端（Windows 中的命令行窗口），输入 python --version 命令，该命令会输出已经安装的 Python 的版本信息。

```
C:\Anaconda3>python --version
Python 3.4.1 :: Anaconda 2.1.0 (64-bit)
```

如上所示，显示了 Python 3.4.1（根据实际安装的版本，版本号可能不同），则说明已正确安装 Python 3.×。接着输入 python，启动 Python 解释器。

```
C:\Anaconda3>python
Python 3.4.1 |Anaconda 2.1.0 (64-bit)| (default, Sep 24 2014, 18:32:42) [MSC v.1600 64 bit (AMD64)] on win32
Type "help", "copyright", "credits" or "license" for more information.
>>>
```

1.1.2　NLTK 的安装

首先，下载 https://gitee.com/huguobn/nltk_data/repository/archive/master.zip 压缩文件。然后，安装本书所需的数据。

一旦数据下载到计算机，便可以使用 Python 解释器进行加载。在 Python 提示符后>>>输入 from nltk.book import *，从 NLTK 的 book 模块加载数据。这个 book 模块包含阅读本书所需的数据，任何时候我们想要找到这些文本，只需要在 Python 提示符后输入它们的名字即可，例如 text1：

```
>>> from nltk.book import *
>>> text1
<Text: Moby Dick by Herman Melville 1851>
```

```
C:\Anaconda2\Lib\site-packages\nltk\book.py
sys.path.append(os.pardir)
```

1.1.3 搜索文本

除了阅读文本之外，还有很多方法可以用于研究文本内容，如词语索引视图会显示一个指定单词的每一次出现（连同部分上下文一起显示）。搜索词 monstrous：

```
from nltk.book import *
text1.concordance("monstrous")
Displaying 11 of 11 matches:
ing Scenes . In connexion with the monstrous pictures of whales ,
I am strongly
of Whale - Bones ; for Whales of a monstrous size are oftentimes
cast up dead u
```

注意：事先要手工解压 C:\nltk_data\corpora\genesis.zip 等压缩文件。词语索引让我们可以看到词的上下文，例如，monstrous 出现的上下文 the __ pictures 和 a __ size。那么，还有哪些词出现在相似的上下文中呢？可以通过在查询的文本名后添加函数名 similar()，然后在括号中插入相关的词来查找。

```
text1.similar("monstrous")
true contemptible christian abundant few part mean careful puzzled
mystifying passing curious loving wise doleful gamesome singular
delightfully perilous fearless
text2.similar("monstrous")
very so exceedingly heartily a as good great extremely remarkably
sweet vast amazingly
```

观察从不同的文本中得到的结果可以发现，text2 中使用的这些词与 text1 完全不同，在 text2 中，monstrous 是正面的意思，如 good、great、sweet 等，有时它的功能像词 very 一样，作为强调成分。

函数 common_contexts 主要研究两个或两个以上词共同的上下文，如 monstrous 和 very。

```
text2.common_contexts(["monstrous", "very"])
a_pretty am_glad a_lucky is_pretty be_glad
```

使用 similar() 和 common_contexts() 函数并比较它们在两个不同文本中的用法。

```
text1.similar("love")
sea man it ship by him hand them whale view ships land me life death
water way head nature fear
text2.similar("love")
```

```
affection sister heart mother time see town life it dear elinor
marianne me word family her him do regard head
text1.common_contexts(["love", "like"])
i_to the_of
```

自动检测出现在文本中特定的词,显示同样上下文中出现的一些词,这只是一个方面。我们也可以判断词在文本中的位置,从文本开头算起在它前面有多少词,这个位置信息可以用离散图表示,每一个竖线代表一个单词,每一行代表整个文本。图 1-1 所示为一些显著的词语用法模式,可以使用下面的方法画出这幅图。

```
text4.dispersion_plot(["citizens", "democracy", "freedom", "duties",
"America"])
```

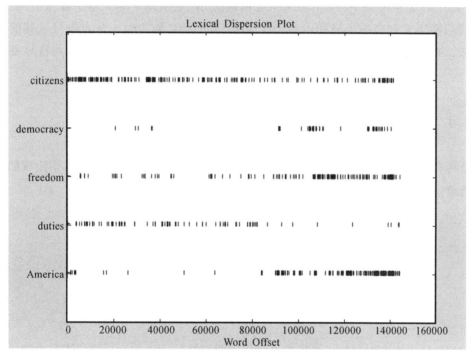

图 1-1　一些显著的词语用法模式

为了更好地理解 dispersion_plot()函数,修改文本进行测试。

```
from nltk.text import Text
text10='It is easy to get our hands on millions of words of text .'
text10=text10.split()
text10=Text(text10)
text10.dispersion_plot(["It", "is", "of"])
```

可以看到,在"It is easy to get our hands on millions of words of text."文本中,从文本开

头算起，在 It 前面有 0 个词，is 前面有 1 个词，of 前面有 9、11 个词（在整个文本中 of 出现了两次）。

下面以刚才看到的不同风格产生一些随机文本。

```
>>>text4.generate()
import nltk
text10='It is easy to get our hands on millions of words of text.'
text10=text10.split()
text10=nltk.text.Text(text10)
text10.generate()
```

注意：第一次运行此命令时，要搜集词序列的统计信息故而执行速度较慢。每次运行该命令，输出的文本都会不同。虽然文本是随机的，但它重用了源文本中常见的词和短语，从而使我们能感觉到它的风格和内容。在 generate()产生输出时，标点符号从其前面的词分割出去，虽然这不是正确的英文格式，但这么做是为了弄清楚文字和标点符号是彼此独立的。

1.1.4 词汇计数

本节将介绍如何使用计算机以各种有用的方式计数词汇。首先，以文本中出现的词和标点符号为单位算出文本从头到尾的长度，使用函数 len()获取长度。

```
from nltk.book import *
len(text3) #44764 个词
len(["to", "be", "or", "not", "to", "be"]) #6 个词
len("to be or not to be") #18 个字符
len(set(w.lower() for w in text3 if w.isalpha()))
```

text3 有 44 764 个词和标点符号或者叫作"标识符"，如"to"。一个标识符表示一个想要放在一组对待的字符序列。当统计文本中标识符的个数时，如 to be or not to be 这句话，可以统计这些词出现的次数（词频），例句中 to 和 be 各出现了 2 次，or 和 not 各出现了 1 次，一共 6 个词。例句中只有 4 个不同的词，text3 中有多少个不同的词呢？一个文本词汇表只是它用到的标识符的集合，因为在集合中所有重复的元素都只算一个，Python 中可以使用命令 set(text3)获得 text3 的词汇表。

```
sorted(set(text3))
len(set(text3)) #2789 个不同的词
len(set(["to", "be", "or", "not", "to", "be"])) #4 个不同的词
```

```
len(set(w.lower() for w in text3 if w.isalpha())) #不区分大小写,由
字母组成,2615
```

用 sorted()得到一个词汇项的排序表（规范化），这个表以各种标点符号开始，然后以 A 开头的词汇，大写单词排在小写单词前面。可以通过求集合中项目的个数间接获得词汇表的大小，再次使用 len()来获得这个数值。尽管 text3 中有 44 764 个标识符，但只有 2 789 个不同的词汇或"词类型"。一个词类型是指一个词在一个文本中独一无二的出现形式或拼写，也就是说，这个词在词汇表中是唯一的，计数的 2 789 个项目中包括标点符号，所以把这些叫作唯一项目类型而不是词类型。对文本词汇丰富度进行计算，对于词类型，用词类型平均使用的次数来度量。每个词类型平均使用的次数越少，文本词汇越丰富。

```
len(text4)/len(set(text4)) #14.941049825712529
len(text5)/len(set(text5)) #7.420046158918563
```

对于特定的词，可以用百分比来度量。接下来，专注于特定的词，计数一个词在文本中出现的次数，计算一个特定的词在文本中占据的百分比。一个特定的词在文本中占据的百分比越小，文本词汇越丰富。

```
100*text4.count('a')/len(text4) #1.46%
100*text5.count('a')/len(text5) #1.26%
```

如果想要对几个文本重复这些计算，可以自己定义一个函数。下面的例子演示了如何定义 lexical_diversity()和 percentage()这两个新的函数。

```
from nltk.book import *
def lexical_diversity(text):
    return len(text)/len(set(text))
def percentage(token, text):
    return 100*text.count(token)/len(text)
```

lexical_diversity()的定义中，我们指定了一个 text 参数，这个参数是想要计算词汇多样性的实际文本的一个"占位符"，即形参，第 2 行是使用这个函数时想要重现的代码段。类似地，percentage()定义了 count 和 total 两个参数。

```
lexical_diversity(text4) #14.941049825712529
lexical_diversity(text5) #7.4200461589185629
percentage('a', text4) #1.46%
print('%.2f%%' % percentage('a', text5)) #1.26%
```

计算结果表明，text5 词汇比 text4 丰富。在本章中，已经遇到了几个函数，如 len()、set()、sorted()和 count()。通常我们会在函数名后面加一对空括号，像 len()函数中的那样，这只是为了表明这是一个函数而不是其他的 Python 表达式。

1.2 文本的表示

1.2.1 列表

文本不外乎是词和标点符号的序列。下面介绍如何在 Python 中表示文本。

```
sent1=['Call', 'me', 'Lys', '.']
' '.join(sent1)
'Call me Lys .'.split() #['Call', 'me', 'Lys', '.']
' '.join(['Call', 'me', 'Lys', '.']) #'Call me Lys .'
```

这个方括号内的内容在 Python 中叫作列表，它就是存储文本的方式，可以查阅它和它的长度。

```
sent1 #['Call', 'me', 'Lys', '.']
len(sent1) #4
```

通过输入名字、等号和一个词列表，组建一些句子：

```
ex1=['Monty', 'Python', 'and', 'the', 'Holy', 'Grail']
sorted(ex1) #['Grail', 'Holy', 'Monty', 'Python', 'and', 'the'], ex1 本身不变。
ex1.sort()
len(set(ex1)) #6
ex1.count('the') #1
```

可以对列表使用 Python 加法运算，这种加法的特殊用途叫作连接，它将多个列表组合为一个列表，包括第一个列表的全部，后面跟着第二个列表的全部。可以把句子连接起来组成一个文本。

```
['Monty', 'Python']+['and', 'the', 'Holy', 'Grail']
['Monty', 'Python'].__add__(['and', 'the', 'Holy', 'Grail'])
['Monty', 'Python'].extend(['and', 'the', 'Holy', 'Grail'])
sent1+sent1
```

当对一个列表使用 append() 时，列表自身会随着操作而更新。

```
sent1.append("Some") #表尾插入
sent1.insert(len(sent1), 'Tail') #表尾插入
sys.path.append(os.pardir)
```

Python 乘法运算可应用于列表，当输入 ['Monty', 'Python']*2 时，会输出 ['Monty', 'Python', 'Monty', 'Python'], ['Monty', 'Python']+['Monty', 'Python']。

1.2.2 索引列表

Python 中的文本是一个词列表,可以通过元素在列表中出现的位置找出一个 Python 列表的元素。表示这个位置的数字叫作这个元素的索引,在文本名称后面的方括号里写下索引,Python 就会找出文本中这个索引处的元素,也可以反过来做,找出一个词第一次出现的索引。

```
text1='Hello , world !'.split()
text1
['Hello', ',', 'world', '!']
len(text1) #4
text1.count('world') #1
text1[2] #'world'
text1.index('world') #2
```

索引是一种用来获取文本中词汇的方式。Python 可以获取子列表,从大文本中任意抽取语言片段便叫作切片。

```
sent=['word1', 'word2', 'word3', 'word4', 'word5', 'word6', 'word7', 'word8', 'word9', 'word10']
sent[0] #'word1'
sent[9] #'word10'
```

注意:索引从 0 开始,第 0 个元素写作 sent[0],其实是第 1 个词"word1";而句子的第 9 个元素是"word10",原因很简单,Python 从计算机内存中的列表获取内容时,要告诉它向前多少个元素。因此,向前 0 个元素使它留在第 1 个元素上。如果我们使用的索引过大就会得到一个错误。

```
sent[10]
Traceback (most recent call last):
  File "<stdin>", line 1, in <module>
IndexError: list index out of range
```

这次不是一个语法错误,因为程序片段在语法上是正确的,它是一个运行时错误,它会产生一个回溯消息,显示错误的上下文、错误的名称 IndexError 及简要的解释说明。

让我们看看切片,这里会发现切片 5:8 包含索引 5、6 和 7 的句子元素。

```
sent[5:8] #['word6', 'word7', 'word8']
sent[5] #'word6'
```

按照惯例,m:n 表示元素 m,…,n-1,[m, n)。正如下一个例子显示的那样,如果切片从列表第一个元素开始,可以省略第一个数字;如果切片到列表最后一个元素处结尾,可以省略第二个数字。

```
sent[:3] #['word1', 'word2', 'word3']
```

可以通过指定它的索引值来修改列表中的元素，也可以用新内容替换掉整个片段。报错的原因是这个列表只有 4 个元素而要获取第 4 个元素后面的元素。

```
sent[0]='First'
sent[9]='Last'
len(sent)  #10
sent[1:9]=['Second', 'Third'] #词组(切片)
sent #['First', 'Second', 'Third', 'Last']
sent[9]
Traceback (most recent call last):
  File "<stdin>", line 1, in <module>
IndexError: list index out of range
```

1.2.3 变　量

定义一个变量 my_sent：

```
my_sent='Hello , world !'.split()
my_sent #['Hello', ',', 'world', '!']
```

变量的名字可以是任何名字，如 my_sent，但变量必须以字母开头，可以包含数字和下划线，最好是选择有意义的变量名。它能提醒代码的含义，也帮助别人读懂 Python 代码。唯一的限制是变量名不能是 Python 的保留字，如 def、if 等。如果使用了保留字，则 Python 会产生一个语法错误。

```
def=['Hello', ',', 'world', '!']
  File "<stdin>", line 1
    def=['Hello', ',', 'world', '!']
       ^
SyntaxError: invalid syntax
```

我们经常会使用变量来保存计算的中间结果，这样做会使代码更容易被读懂，因此，len(set(my_sent))也可以写作：

```
vocab=set(my_sent)
vocab_size=len(vocab)
vocab_size #4
```

注意：名称是大小写敏感的，变量名不能包含空格，但可以用下划线把单词分开，不要插入连字符来代替下划线，因为 Python 会把连字符解释为减号。

1.2.4 字符串

可以把一个字符串指定给一个变量，索引一个字符串（取元素），切片一个字符串（求子串）。

```
s='Hello'
type(s) #<type 'str'>
s[0] #'H'
s.index('H') #0
s.find('H') #0
s[:4] #'Hell'
```

还可以对字符串执行乘法（复制）和加法（连接）。

```
s*2 #'HelloHello'
s+s #'HelloHello'
s+' !' #'Hello !'
```

可以把词用列表连接起来组成单个字符串，或者把字符串分割成一个列表。

```
'Hello'+' '+'World' #'Hello World'
'Hello World'.split() #['Hello', 'World']
' '.join(['Hello', 'World']) #'Hello World'
'Hello' + ' ' + 'World' #'Hello World'
```

1.3 语言计算

在本节中，我们将重新拾起是什么让一个文本不同于其他文本这样的问题，并使用程序自动寻找特征词汇和文字表达。

```
saying=['After', 'all', 'is', 'said', 'and', 'done', ',', 'more',
'is', 'said', 'than', 'done']
' '.join(saying)
tokens=set(w.lower() for w in saying) #After 和 after 重复
tokens
tokens=sorted(tokens) #排序,规范化
tokens
tokens[-2:] #切片,取最后2个单词
```

1.3.1 频率分布

如何能自动识别文本中最能体现文本的主题和风格的词汇？试想一下，要找到一本书中使用最频繁的 5 个词你会怎么做？一种方法是为每个词项设置一个计数器，计数器可能需要几千行代码，这将是一个极其烦琐的过程，如此烦琐以至于我们宁愿把任务交给机器来做。

频率分布,将告诉我们在文本中每一个词项的频率,以及文本中词的总数是如何分布在词项中的。由于经常需要在语言处理中使用频率分布,NLTK 中内置了它。下面使用 FreqDist 寻找 text1 中最常见的 5 个词。

```
from nltk.book import *
from nltk import FreqDist
fdist1=FreqDist(text1)
fdist1
#FreqDist({',': 18713, 'the': 13721, '.': 6862, 'of': 6536, 'and': 6024, 'a': 4569, 'to': 4542, ';': 4072, 'in': 3916, 'that': 2982, ...})
fdist1.most_common()[:5]
#[(',', 18713), ('the', 13721), ('.', 6862), ('of', 6536), ('and', 6024)]
fdist1['whale'] #906
fdist1.N() #260819
```

260 819 个词中,whale 分布了 906 个。第一次调用 FreqDist 时,传递文本的名称作为参数,可以看到已经被计算出来的 text1 中总的词数 fdist1.N(),表达式 keys()提供了文本中所有不同类型的列表,可以通过切片 fdist1.most_common()[:5]查看这个列表的前 5 项。

```
from nltk import FreqDist
text10='I love you , you love me , we love each other . So English is easy !'.split()
fdist1=FreqDist(text10)
fdist1.items() #dict(fdist1)
#dict_items([('other', 1), ('.', 1), ('me', 1), (',', 2), ('English', 1), ('I', 1), ('So', 1), ('!', 1), ('each', 1), ('easy', 1), ('you', 2), ('we', 1), ('is', 1), ('love', 3)])
fdist1.keys()
#dict_keys(['other', '.', 'me', ',', 'English', 'I', 'So', '!', 'each', 'easy', 'you', 'we', 'is', 'love'])
fdist1.most_common()[:5]
[('love', 3), (',', 2), ('you', 2), ('other', 1), ('.', 1)]
fdist1['love'] #3
fdist1.plot(5, cumulative=True)
fdist1.N() #18
fdist1.hapaxes() #['other', '.', 'me', 'English', 'I', 'So', '!', 'each', 'easy', 'we', 'is']
```

使用 fdist1.plot(3, cumulative= True)可以产生一个词汇的累积频率图,如图 1-2 所示。

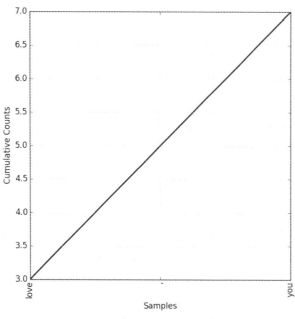

图 1-2 最常用词的累积频率图

1.3.2 高频长词

让我们来看看文本中的长词,也许它们有更多的特征和信息量。下面找出文本词汇表长度中超过 7 个字符的词,定义这个特征为 P,则 P(w)为真,当且仅当词 w 的长度大于 7 个字符。

```
{w | w∈V & P(w)}
[w for w in V if P(w)]
```

注意:它产生一个列表,而不是集合,这意味着可能会有相同的元素。

```
text1="I am asking you to give these victims a little bit help, even
if it's just one penny. I would still appreciate that. I have learned
to appreciate myself more, both my body and the kind of person that I
am . The next day, we reluctantly left the Dalian, this trip I not only
appreciate the beautiful scenery of the sea, but also to understand some
reason, I am really glad that ah! ".split()
text1=[w.lower() for w in text1 if w.isalpha()]
V=set(text1)
long_words=[w for w in V if len(w)>7]
sorted(long_words)  #['appreciate', 'beautiful', 'reluctantly',
'understand']
```

对于词汇表 V 中的每一个词 w,检查 len(w)是否大于 7,其他词汇将被忽略。以下是所有长度超过 7 个字符出现次数超过 2 次的词。

```
from nltk import FreqDist
fdist1=FreqDist(text1)
sorted(w for w in set(text1) if len(w)>7 and fdist1[w]>2)
#['appreciate']
```

len(w)>7 保证所有长度超过 7 个字符，fdist1[w]>2 保证出现次数超过 2 次，这样便能成功地自动识别出与文本内容相关的高频词。

1.3.3 词语搭配和双连词

搭配是频繁在一起出现的词序列。一个搭配的特点是其中的词不能被类似的词置换，例如，foreign language 是一个搭配而 the language 不是。文本中出现的搭配很能体现文本的风格，要获取搭配，先要从提取文本中的词对也就是双连词开始，使用函数 bigrams( )实现。

```
from nltk import bigrams
list(bigrams(['more', 'is', 'said', 'than', 'done']))
#[('more', 'is'), ('is', 'said'), ('said', 'than'), ('than', 'done')]
```

在这里可以看到词对 than-done 是一个双连词，在 Python 中写成元组('than', 'done')。搭配基本上就是频繁的双连词，特别地，我们希望找到基于单个词的频率 FreqDist()预期得到的更频繁出现的双连词 collocations()。

```
from nltk.book import *
text1.collocations() #old man; years ago; never mind; one hand
```

1.3.4 词长计数

可以查看文本中词长的分布，通过创建一长串数字的列表，其中每个数字是文本中对应词的长度。

```
from nltk import FreqDist
text10='I love you , you love me , we love each other . So English is easy !'.split()
[len(w) for w in text10]
#[1, 4, 3, 1, 3, 4, 2, 1, 2, 4, 4, 5, 1, 2, 7, 2, 4, 1]
```

#I 和标点符号是不同的词，但长度都是 1，这就是特征。

```
fd=FreqDist(len(w) for w in text10)
print(dict(fd))
#{1: 5, 2: 4, 3: 2, 4: 5, 5: 1, 7: 1}
#长度为 1 的词有 5 个
```

```
fd.keys() #dict_keys([1, 2, 3, 4, 5, 7])
fd #FreqDist({1: 5, 4: 5, 2: 4, 3: 2, 5: 1, 7: 1})
fd.items() #dict_items([(1, 5), (2, 4), (3, 2), (4, 5), (5, 1), (7, 1)])
fd.values() #dict_values([5, 4, 2, 5, 1, 1])
```

以导出 text10 中每个词长度的列表开始，然后 FreqDist 计数列表中每个长度数字出现的次数。结果是一个包含 16 个元素的分布，每一个元素是一个数字，对应文本中一个词标识符，但是只有 6 个不同的元素，从 1 到 7。因为只有 7 个不同的词长，也就是说，有由 1 个字符、2 个字符、…、7 个字符组成的词，而没有由 6 个或更多字符组成的词，有人可能会问不同长度的词的频率是多少？例如，文本中有多少长度为 4 的词？长度为 5 的词是否比长度为 4 的词多？下面我们回答这些问题。

```
fd.items() #dict_items([(1, 5), (2, 4), (3, 2), (4, 5), (5, 1), (7, 1)])
fd.max() #1，词长为 1 的次数为 5，最大
fd[3] #2，词长为 3 的次数为 2
fd.freq(3) #0.1111111111111111=2/18
fd[3]/fd.N() #0.1111111111111111=2/18
fd.plot(cumulative=True)
```

由此可以看到，最频繁的词长度是 1 和 4，长度为 3 的词有 2 个，约占全部词汇的 11%，关于词长的进一步分析可以帮助我们了解作者、文体或语言之间的差异。表 1-1 总结了 nltk 频率分布类中定义的函数。

表 1-1 nltk 频率分布类 FreqDist 中定义的函数

例子	描述
fdist=FreqDist(samples)	创建包含给定样本的频率分布
fdist[3]	计数词长为 3 的次数
fdist.freq(3)	词长为 3 的频率
fdist.N()	样本总数
fdist.keys()	样本列表
fdist.max()	数值最大的样本
fdist.plot(cumulative=True)	绘制累积频率分布图

1.4 流程控制

1.4.1 条　件

Python 支持多种运算符。数值比较运算符如表 1-2 所示。

表 1-2 数值比较运算符

运算符	关系
<	小于
<=	小于等于
==	等于
!=	不等于
>	大于
>=	大于等于

可以使用这些运算符从句子中选出不同的词。

```
from nltk.book import *
sent1
[w for w in sent1 if len(w)<4] #选择词长小于4的词
```

这个例子的模式为[w for w in text if condition]，其中 condition 是 Python 中的一个条件，集合描述法：{w|w ∈ sent1^len(w)<4}，SQL 表示：select w from sent1 where len(w)<4。在前面的例子中，条件是数值比较，也可以使用表 1-3 中列出的函数测试词汇的属性。

表 1-3 测试函数

函数	含义
s.startswith(t)	测试 s 是否以 t 开头，注意不是 begins
s.endswith(t)	测试 s 是否以 t 结尾
t in s	测试 s 是否包含 t（子串），与 set(t)<set(s)不同（子集）
s.islower()	测试 s 中所有字符是否都是小写字母，s.lower()
s.isupper()	测试 s 中所有字符是否都是大写字母
s.isalpha()	测试 s 中所有字符是否都是字母
s.isalnum()	测试 s 中所有字符是否都是字母或数字
s.isdigit()	测试 s 中所有字符是否都是数字
s.istitle()	测试 s 是否首字母大写

下面是一些用来从文本中选择词汇的例子。

```
sorted(w for w in set(text1) if w.endswith('ableness')) #选择以
-ableness 结尾的词
#['comfortableness',  'honourableness',  'immutableness',
'indispensableness', ...]
sorted(item for item in set(text6) if item.istitle())
#['A', 'Aaaaaaaaah', 'Aaaaaaah', 'Aaaaaah', 'Aaaah', 'Aaaaugh',
'Aaagh', ...]#首字母大写的词
sorted(item for item in set(sent7) if item.isdigit())
#完全由数字组成的词，['29', '61']
```

还可以构造更复杂的条件。如果 c 是一个条件，那么 not c 也是一个条件。如果有两个条件 c1 和 c2，那么可以使用合取和析取将它们合并形成一个新的条件 c1 and c2 以及 c1 or c2。

运行下面的例子，尝试解释每一条指令中所发生的事情。然后，试着自己组合一些条件。

```
sorted(w for w in set(text7) if '-' in w and 'index' in w) #选择
包含-和 index 的词
text10='I love you , you love me , we love each other . So English is easy !'.split()
sorted(t for t in set(text10) if 'er' in t or 're' in t) #选择包含 er 或者 re 的词['other']
```

1.4.2 列表推导

```
text1='This is a book'.split() #[len(w) for w in text1] #[4, 2, 1, 4]
[w.upper() for w in text1] #['THIS', 'IS', 'A', 'BOOK']
```

这些表达式形式为[f(w) for …]或[w.f() for …]，其中 f 是一个函数，这里用来计算词长或把字母转换为大写。现阶段还不需要理解两种表示方法 f(w)与 w.f()之间的差异，只需学习对列表所有元素执行相同操作的这种 Python 习惯用法。在前面的例子中，遍历 text1 中的每一个词，一个接一个地赋值给变量 w 并在变量上执行指定的操作，上面描述的表示法被称为列表推导。回到计数词汇的问题，这里使用相同的习惯用法。

```
len(set(w.lower() for w in text1))
```

不重复计算像 This 和 this 这样仅仅大小写不同的词，就可以从词汇表计数中抹去了 2 000 个，还可以更进一步，通过过滤掉所有非字母元素，从词汇表中消除数字和标点符号。

```
len(set(w.lower() for w in text1 if w.isalpha()))
```

这个例子稍微有些复杂，将所有纯字母组成的词小写去重。也许只计数小写的词会更简单一些，但这却是一个错误的答案，为什么呢？This is a book 计数得到 3 个词，This 不是小写的词。

```
text1='This is a book'.split()
len(set(w for w in text1 if w.islower())) #3, ×
len(set(w.lower() for w in text1 if w.isalpha())) #4, √
```

1.4.3 控制结构

大多数编程语言允许在条件表达式或者说 if 语句条件满足时执行代码块。在下面的程序中，创建一个叫 word 的变量包含字符串值'cat'，在 if 语句中检查 len(word)<5 是否为真，cat 的长度确实<5，所以 if 语句下的代码块被调用，print 语句被执行，向用户显示一条消息。

```
word='cat'
if len(word)<5:
    print('word length is less than 5')

#word length is less than 5
len(word)  #3
```

如果改变测试条件为 len(word)>=5，检查词的长度是否大于或等于 5，那么测试将为假，此时，if 语句后面的代码段将不会被执行，没有消息显示给用户。

```
if len(word)>=5:
    print 'word length is greater than or equal to 5'
```

另一个控制结构是 for 循环。

```
for word in ['Call', 'me', 'LYS', '.']:
    print(word)
```

Python 以循环的方式执行里面的代码，用 word 作为循环控制变量。

1.4.4 条件循环

将 if 语句和 for 语句结合，遍历列表中每一项，只输出结尾字母是 l 的词。

```
sent1=['Call', 'me', 'LYS', '.']

for w in sent1:
    if w.endswith('l'):
        print(w)  #Call

[w for w in sent1 if w.endswith('l')]  #Call
```

也可以指定当 if 语句的条件不满足时采取的行动。

```
sent1=['Call', 'me', 'LYS', '.']
for w in sent1:
    if w.islower():
        print(w, ' is a lowercase word')
    elif w.istitle():
        print(w, ' is a titlecase word')
    else:
        print(w, 'is punctuation')  #标点

Call is a titlecase word
```

```
me is a lowercase word
LYS is punctuation #LYS 不是标点符号！×
. is punctuation
```

最后，把一直在探索的习惯用法组合起来，创建一个包含 cie 或者 cei 词的列表，循环输出其中的每一项。

```
text2='I am reading , I am writing . I played table tennis yesterday .'
ws=sorted(w for w in set(text2.split()) if 'ing' in w or 'ed' in w)
for w in ws:
    print(w)
```

小　结

（1）在 Python 中文本用列表来表示，可以使用索引、分片和 len()函数对列表进行操作。

（2）token 是指文本中给定词的特定出现，type 则是指词作为一个特定序列字母的唯一形式(This，this)。使用 len(text)计数 token，使用 len(set(text))计数 type。

（3）使用 sorted(set(t))获得文本 t 的词汇表。

（4）使用[f(x) for x in text]对文本的每一项目进行操作。

（5）为了获得没有大小写区分和忽略标点符号的词汇表，可以使用 set(w.lower() for w in text if w.isalpha())、islower()。

（6）使用 for 语句对文本中的每个词进行处理，如 x.f() for w in t。

（7）使用 if 语句测试一个条件，if len(word)<5。

（8）频率分布是项目连同频率计数的集合，例如，一个文本中词与它们出现的频率。

（9）函数是指定了名字并且可以重用的代码块，函数通过 def 关键字定义，例如，在 def mul(x, y)中 x 和 y 是函数的参数，起到实际数据值占位符的作用。

（10）函数通过指定它的名字及一个或多个放在括号里的实参来调用，就像 mul(3, 4)或者 len(text1)一样。

练　习

1. 尝试使用 Python 解释器作为一个计算器，输入表达式，如 12/(4+1)。

```
C:\Anaconda3>python
Python 3.4.1 |Anaconda 2.1.0 (64-bit)| (default, Sep 24 2014, 18:32:42) [MSC v.1600 64 bit (AMD64)] on win32
Type "help", "copyright", "credits" or "license" for more information.
>>> 12/(4+1)
```

```
2.4
>>> float(12)/(4+1)  #注意整除和浮点除。
2.4
>>> 12.0/(4+1)
2.4
>>> 1.0*12/(4+1)
2.4
```

2. 26 个字母可以组成 26^{10} 个 10 字母长的字符串，也就是 141167095653376L（结尾处的 L 只是表示这是 Python 长数字格式），试问 100 个字母长度的字符串可能有多少个？Python 支持大整数运算。

3. Python 乘法运算可应用于链表。当输入['Monty', 'Python']*20 或者 3*sent1 时，会发生什么？

4. 复习 1.1 节关于语言计算的内容，在 text2 中有多少个词？有多少个不同的词？

```
from nltk.bo import *
len(text2) #141576
len(set(text2)) #6833
len(set(w.lower() for w in text2)) #6403
```

5. 比较表 1-4 中幽默和言情小说的词汇多样性得分，哪一个文体中词汇更丰富？

```
from nltk.corpus import brown
brown.words()
brown.categories()
['adventure', 'belles_lettres', 'editorial', 'fiction', 'government',
'hobbies', 'humor', 'learned', 'lore', 'mystery', 'news', 'religion',
'reviews', 'romance', 'science_fiction'], rumor
from __future__ import division
tens=len(brown.words(categories='humor')) #82345
vocabulary=len(set(brown.words(categories='romance'))) #11935
diversity=tens/vocabulary
```

表 1-4 Brown 语料库中各种文体的词汇多样性

体裁	标识符	类型	词汇多样性
技能和爱好	82345	11935	6.9
幽默	21695	5017	4.3
小说：科学	14470	3233	4.5
新闻：报告文学	100554	14394	7.0
小说：浪漫	70022	8452	8.3
宗教	39399	6373	6.2

6. 制作《理智与情感》中四个主角(Sense and Sensitivity)Elinor、Marianne、Edward 和 Willoughby 的分布图。在这部小说中根据男性和女性所扮演的不同角色，你能观察到什么？你能找出一对夫妻吗？

https://blog.csdn.net/weixin_40122035/article/details/82380871

text2.dispersion_plot(['Elinor', 'Edward', 'Willoughby', 'Marianne'])

7. 查找 text5 中的搭配。

```
from nltk import *
bg=bigrams(text5)
fd=FreqDist(bg)
fd.keys()[:300]
from nltk.bo import *
text5.collocations()
Building collocations list
wanna chat; PART JOIN; MODE #14-19teens; JOIN PART; cute.-ass MP3;
MP3 player; times ...; ACTION watches; guys wanna; song lasts; last
night; ACTION sits; -...)...-S.M.R.; Lime Player; Player 12%; dont
know; lez gurls; long time; gently kisses; Last seen
collocations[ˌkɒləʊˈkeɪʃənz]n.词组；组合；(词语的) 搭配，组配；[例句]"strong tea" and "by accident" are English collocations.strong tea 与 by accident 是英语经常搭配的词组。
```

8. 思考下面的 Python 表达式：len(set(text4))，并说明这个表达式的用途。简要描述在执行此计算中涉及的两个步骤。

9. 复习 1.2 节关于列表和字符串的内容。

（1）定义一个字符串，并且将它分配给一个变量，如 my_string='My String'。用两种方法输出这个变量的内容，一种是通过简单地输入变量的名称，然后按回车；另一种是通过使用 print 语句。思考它们有何不同。

```
my_string='My String'
my_string
print my_string
```

（2）尝试使用 my_string+my_string 或者用它乘以一个数将字符串添加到它自身，如 my_string*3。注意：连接在一起的字符串之间没有空格。怎样能解决这个问题？

```
my_string+my_string
my_string+' '+my_string
(my_string+' ')*3 #???
```

10. 使用语法 my_sent=["My", "sent"]，定义一个词列表变量 my_sent。

（1）使用' '.join(my_sent)将其转换成一个字符串。

（2）使用 split() 在指定的地方将字符串分割回列表。

```
my_sent=["My", "sent"]
' '.join(my_sent)
my_sent="My sent"
my_sent.split()
```

11. 定义几个包含词列表的变量，如 phrase1、phrase2 等，将它们连接在一起组成不同的组合（使用加法运算符），最终形成完整的句子。len(phrase1+phrase2) 与 len(phrase1)+len(phrase2) 之间的关系是什么？

```
phrase1=['Good', 'morning']
phrase2=['July', 'Tree']
phrase1+phrase2
len(phrase1+phrase2) #4
len(phrase1)+len(phrase2) #4
#len(phrase1+phrase2)=len(phrase1)+len(phrase2)
```

12. 考虑下面两个具有相同值的表达式。哪一个在 NLP 中更常用？为什么？

（1）"Monty Python"[6:12]
（2）["Monty", "Python"][1]

13. 我们已经看到如何用词列表表示一个句子，其中每个词是一个字符序列。sent1[2][2] 代表什么意思？为什么？请用其他的索引值做实验。

```
sent1=['She', 'sells', 'sea', 'shell']
sent1[2] #'sea'
sent1[2][2] #'a'
#sent1 中的第三个单词的第三个字母
```

14. 在变量 sent3 中保存的是 text3 的第一句话。

在 sent3 中 the 的索引值是 1，因为 sent3[1] 的值是"the"。sent3 中"the"其他出现的索引值是多少？

```
1, 5, 8
from nltk.book import *
for i in range(len(sent3)-2):
    if sent3[i]=='the':
        print i


>>> from nltk.book import *
*** Introductory Examples for the NLTK Book ***
```

```
Loading text1, ..., text9 and sent1, ..., sent9
Type the name of the text or sentence to view it.
Type: 'texts()' or 'sents()' to list the materials.
text1: Moby Dick by Herman Melville 1851
text2: Sense and Sensibility by Jane Austen 1811
text3: The Book of Genesis
text4: Inaugural Address Corpus
text5: Chat Corpus
text6: Monty Python and the Holy Grail
text7: Wall Street Journal
text8: Personals Corpus
text9: The Man Who Was Thursday by G . K . Chesterton 1908
>>> i=0
>>> while i<len(sent3):
...     try:
...         i=sent3.index('the', i+1)
...     except ValueError:
...         break
...     else:
...         print i
...
1
5
8

from nltk.book import *
[i for i in range(len(sent3)-2) if sent3[i]=='the']
```

15. 复习 1.4 节讨论的条件语句。在聊天语料库(text5)中查找所有以字母 b 开头的词，并按字母顺序显示出来。

```
from nltk.bo import *
sorted([w for w in text5 if w.startswith('b')])
sorted([w for w in set(text5) if w.startswith('b')])
text5.findall('<b.*>') #????
```

16. 在 Python 解释器提示符下输入表达式 range(10)。

再尝试 range(10, 20)、range(10, 20, 2)和 range(20, 10, -2)。在后续章节中将看到这个内置函数的多种用途。

```
range(10)
range(10, 20)
range(10, 20, 2)
range(20, 10, -2)
for i in range(10, 20, 2):
    print(i)
```

17. 使用 text9.index() 查找词 sunset 的索引值，需要将这个词作为一个参数插入圆括号之间。找到完整的句子中包含这个词的切片。

```
from nltk.bo import *
text9.index('sunset') #629
text9[629-25:629+25]
from nltk.bo import *
def find_word(text, word):
    pos=0
    while pos<len(text):
        try:
            pos=text.index(word, pos)+1
            print pos
        except Exception as e:
            print 'all have been found!'
            return

find_word(list(text9), 'sunset')
```

18. 使用列表加法、set 和 sorted 操作，计算句子 sent1…sent8 的词汇表。

```
from nltk.bo import *
sorted(set(sent1).union(set(sent2)))
sorted(set(sent1+sent2+sent3+sent4+sent5+sent6+sent7+sent8))
sent1=['a', 'b']
sent2=['a', 'b', 'c']
```

19. 下面两行之间的差异是什么？哪一个的值较大？其他文本也是同样情况吗？

```
sorted(set([w.lower() for w in text1]))
sorted([w.lower() for w in set(text1)])
len(sorted(set([w.lower() for w in text1]))) #17231
len(sorted([w.lower() for w in set(text1)])) #19317
#第二个更大，第二个的值应大于等于第一个的值，因为在第二个中大小写不同的单词都会被保存下来。
```

```
text1=['a', 'b', 'A', 'B']
sorted(set([w.lower() for w in text1])) #
sorted([w.lower() for w in set(text1)])
```

20. w.isupper()和not w.islower()这两个测试之间的差异是什么？

```
sent=['She', 'sells', 'sea', 'shells', 'by', 'the', 'sea', 'shore']
[w for w in sent if w.isupper()]
[w for w in sent if not w.islower()]
```

w.isupper()返回的是 w 是否为全大写的字母，not w.islower()返回的是 w 是否全不是小写字母（可能包含数字等）。

21. 写一个切片表达式提取 text2 中最后两个词。

```
from nltk.bo import *
text2[-2:]
```

22. 找出聊天语料库(text5)中所有四个字母的词。

使用频率分布函数(FreqDist)，以频率从高到低显示这些词。

```
from nltk.bo import *
from nltk import *
fd=FreqDist([w for w in text5 if len(w)==4])
print fd
<FreqDist: 'JOIN': 1021, 'PART': 1016, 'that': 274, 'what': 183,
'here': 181, '.
...': 170, 'have': 164, 'like': 156, 'with': 152, 'chat': 142, ...>
```

23. 复习 1.4 节中条件循环的讨论。使用 for 和 if 语句组合循环遍历《巨蟒和圣杯》(text6)电影剧本中的词，输出所有的大写词，每行输出一个。

```
from nltk.bo import *
[w for w in text6 if w.isupper()]
from nltk.bo import *
for w in text6:
    if w.isupper():
        print w
```

24. 写表达式找出 text6 中所有符合下列条件的词。
结果应该是词列表的形式：['word1', 'word2', …]。
（1）以 ize 结尾；
（2）包含字母 z；
（3）包含字母序列 pt；
（4）除了首字母外全部是小写字母的词（即 titlecase）。

```
from nltk.bo import *
[w for w in text6 if w.endswith('ize')] #a
[w for w in text6 if 'z' in w] #b
[w for w in text6 if 'pt' in w] #c
[w for w in text6 if w.istitle()] #d
[w for w in list(text6) if w.endswith('ize') and w.find('pt')!=-1
and w[0].isupper() and w[1:].islower()]
<Text: Monty Python and the Holy Grail>
[w for w in text6 if w.endswith('ment')] #a
['moment', 'government', 'Torment', 'punishment']
import re
[w for w in text6 if re.search(r'ment$', w)] #a
text6.findall(r'<\w*ment>')
```

25. 定义 sent 为词列表['she', 'sells', 'sea', 'shells', 'by', 'the', 'sea', 'shore']。编写代码执行以下任务。

（1）输出所有 sh 开头的单词；

（2）输出所有长度超过 4 个字符的词。

```
sent=['she', 'sells', 'sea', 'shells', 'by', 'the', 'sea', 'shore']
[w for w in sent if w.startswith('sh')]
[w for w in sent if len(w)>4]
```

26. 下面的 Python 代码是做什么的？sum(len(w) for w in text1)，你可以用它来计算出一个文本的平均字长吗？

```
from nltk.bo import *
from __future__ import division
sum(len(w) for w in text1)/len(text1)
#计算 text1 文本中所有单词的总长度
'%.2f' % (sum(len(w) for w in text1)/len(text1))
```

27. 定义一个名为 vocab_size(text)的函数，以文本作为唯一的参数返回文本的词汇量。

```
def vocab_size(text):
    return len(set(w.lower() for w in text if w.isalpha()))

import nltk
text='She sells sea shells by the sea shore'
text=nltk.word_tenize(text)
vocab_size(text)
```

28. 定义一个函数 percent(word，text)，计算一个给定的词在文本中出现的频率，结果以百分比表示。

```
from nltk import *
from __future__ import division
def percent(word, text):
    return 100*text.count(word)/len(text)

def percent2(word, text):
    fd=FreqDist(text)
    return fd.freq(word)*100

text='Excuse me . Is there any summer resort around the town ?'
text='She sells sea shells by the sea shore .'
word='me'
text=word_tenize(text)
percent(word, text)
def percent3(word, text):
    fd=FreqDist(text)
    return '{:.2f}%'.format(fd[word]*100/len(text))
#'%.2f%%' % 0.3333
```

29. 我们一直在使用集合存储词汇表，试试下面的 Python 表达式，set(sent3)<set(text1)，尝试在 set()中使用不同的参数，它是做什么用的？你能想到一个实际的应用吗？

```
from nltk.book import *
set(sent3)<set(text1) #True
#sent3 中的每一个元素是否都在 text1 中
#可用于判断一个集合是否为另一个集合的子集
text1=['the', 'beginning', 'God', 'created', 'the', 'heaven', 'and', 'the', 'earth', '.', 'In']
sent3=['In', 'the', 'beginning']
set(sent3)<set(text1) #True
sent3<text1 #False
sent3 in text1 #True
```

阅读材料

缺失值的处理

在 sklearn 的 preprocessing 包中包含对数据集中缺失值的处理，主要是应用 Imputer 类进行处理。首先需要说明的是，numpy 的数组中可以使用 np.nan、np.NaN(Not A Number)来代替缺失值，对于数组中是否存在 nan 可以使用 np.isnan() 来判定。使用 type(np.nan) 或者 type(np.NaN) 可以发现该值其实属于 float 类型，代码如下：

```
>>> type(np.NaN)
<type 'float'>
>>> type(np.nan)
<type 'float'>
>>> np.NaN
nan
>>> np.nan
nan
```

因此，处理数据集中包含缺失值的一般步骤如下：
（1）使用字符串'nan'来代替数据集中的缺失值；
（2）将该数据集转换为浮点型便可以得到包含 np.nan 的数据集；
（3）使用 sklearn.preprocessing.Imputer 类来处理使用 np.nan 对缺失值进行编码过的数据集。
代码如下：

```
>>> from sklearn.preprocessing import Imputer
>>> imp=Imputer(missing_values='NaN', strategy='mean', axis=0)
>>> X=np.array([
[1, 2],
[np.nan, 3],
[7, 6]])
第一列：
1
np.nan
7
去掉'NaN'：
1
7
按列求均值：
(1+7)/2=4
第二列：
2
3
```

```
6
```
按列求均值：
(2+3+6)/3=3.66666667

```
>>> Y=[
[np.nan, 2],
[6, np.nan],
[7, 6]]
```
第一列的 np.nan 用训练的均值 4 代替
第二列的 np.nan 用训练的均值 3.66666667 代替
```
>>> imp.fit(X)
Imputer(axis=0, copy=True, missing_values='NaN', strategy='mean', verbose=0)
>>> imp.transform(Y)
array([[ 4.        ,  2.        ],
       [ 6.        ,  3.66666667],
       [ 7.        ,  6.        ]])
>>> imp.transform(X)
array([[ 1.,  2.],
       [ 4.,  3.],
       [ 7.,  6.]])
```

上述代码使用数组 X 去"训练"一个 Imputer 类，然后用该类的对象去处理数组 Y 中的缺失值，缺失值的处理方式是使用 X 中的均值（axis=0 表示按列进行）代替 Y 中的缺失值。当然也可以使用 imp 对象来对 X 数组本身进行处理。

通常，我们的数据都保存在文件中，也不一定都是 Numpy 数组生成的，因此缺失值可能不一定是使用 nan 来编码的，对于这种情况可以参考以下代码：

```
>>> line='1, ?'
>>> line=line.replace(', ?', ', nan')
>>> line
'1, nan'
>>> Z=line.split(', ')
>>> Z
['1', 'nan']
>>> Z=np.array(Z, dtype=float)
>>> Z
array([ 1., nan])
>>> imp.transform(Z)
array([[ 1.        ,  3.66666667]])
```

上述代码 line 模拟从文件中读取出来的一行数据，使用 nan 来代替原始数据中的缺失值编码，将其转换为浮点型，然后使用 X 中的均值填补 Z 中的缺失值。

第 2 章 文本语料库获取

2.1 文本语料库概述

2.1.1 gutenberg 语料库

NLTK 包含 gutenberg 项目电子文本档案的一小部分文本，下面用 Python 解释器加载 NLTK 包，尝试使用 nltk.corpus.gutenberg.fileids()。

```
>>> from nltk.corpus import gutenberg
>>> gutenberg.fileids() #gutenberg语料库中的文件标识符
['austen-emma.txt', 'austen-persuasion.txt', ..., 'whitman-leaves.txt']
```

挑选这些文本的第一个 austen-emma.txt 并给它一个简短的名称，然后找出它包含多少个词。

```
>>> emma=gutenberg.words('austen-emma.txt') #word_tokenize()
>>> len(emma)
192427
>>> len(gutenberg.raw('austen-emma.txt').split())
158167
```

可以通过找到它的位置 C:\nltk_data\corpora\gutenberg\austen-emma.txt，把自己的文本放到 C:\nltk_data\corpora\gutenberg\love.txt。

```
#I love you, you love me, we love each other, so English is easy.
love=gutenberg.words('love.txt')
len(love) #18,可见标点符号也算一个词。
len('I love you, you love me, we love each other, so English is easy.'.split()) #14
```

写一个简短的程序，通过遍历前面列出的 gutenberg 文件标识符列表相应的 fileid，然后统计每个文本。

```
from nltk.corpus import gutenberg
for f in gutenberg.fileids():
    c=len(gutenberg.raw(f)) - gutenberg.raw(f).count(' ') #减去空格符
```

```
    w=len(gutenberg.words(f))  #分词
    s=len(gutenberg.sents(f))  #断句
    v=len(set(w.lower() for w in gutenberg.words(f)))  #单词集
    print('%.2f %.2f %.2f %s ' % (c/w, w/s, w/v, f))

3.84 24.82 26.20 austen-emma.txt
3.96 26.20 16.82 austen-persuasion.txt
```

这个程序计算了每个文本的三种统计量：平均词长（字符数）、平均句子长度（单词数）和每个词出现的平均次数（平均词频）。可见，平均词长似乎是英语的一个一般属性，因为它的值总是 3，故通常不将它作为特征。相比之下，平均句子长度和每个词出现的平均次数看上去是作者个人的特点，故通常将它作为特征。

前面的例子表明怎样才能获取原始文本，而不用把它分割成标识符(words())。raw()函数给出没有进行过任何语言学处理的文件的内容，因此 len(gutenberg.raw('love.txt'))表示文本中出现的字符个数，包括词之间的空格，sents()函数把文本划分成句子，其中每一个句子是一个词列表['I', 'love', 'you', '.']。

```
I love you. you love me. we love each other. so English is easy.
#句点（.）后要加一个空格符。
from nltk.corpus import gutenberg
len(gutenberg.raw('love.txt'))  #64，回车换行符，算2个字符
gutenberg.sents('love.txt')  #C:\nltk_data\tokenizers\punkt.zip#
事先解压
#[['I', 'love', 'you', '.'], ['you', 'love', 'me', '.'], ...]
len(gutenberg.sents('love.txt'))  #4
gutenberg #<PlaintextCorpusReader in 'C:\\nltk_data\\corpora\\gutenberg'>
type(gutenberg)  #<class 'nltk.corpus.reader.plaintext.PlaintextCorpusReader'>
```

2.1.2 webtext 和 nps_chat 语料库

```
NLTK 的 webtext 小集合包括 Firefox 交流论坛等内容。
from nltk.corpus import webtext
for f in webtext.fileids():
    print(f, webtext.raw(f)[:6])
firefox.txt Cookie
grail.txt SCENE
```

nps_chat 即时消息聊天会话语料库，例如，10-19-20s_706posts.xml 包含 2006 年 10 月 19 日从聊天室收集的帖子。

2.1.3　brown 语料库

brown 语料库是第一个百万词级的英语电子语料库，包含 500 个不同来源的文本，可按照文体分类，如 news（新闻）、editorial（社论）等，见表 2-1。

表 2-1　brown 语料库示例文档

ID	文件	类别	描述
A16	ca16	news	news Chicago Tribune: Society Reportage
B02	cb02	editorial	editorial Christian Science Monitor: Editorials

可以将语料库作为词列表或者句子列表来访问，每个句子本身也是一个词列表，可以指定特定的类别或文件阅读。

```
from nltk.corpus import brown
brown.categories()
brown.words(categories='news')
brown.words(fileids='ca16')
```

brown 语料库是一个研究文体之间系统性差异的资源。

下面比较不同文体中的情态动词的用法。首先，产生特定文体的计数。

```
from nltk.corpus import brown
from nltk import FreqDist
news_text=brown.words(categories='news')
fdist=FreqDist([w.lower() for w in news_text])
modals=['can', 'could', 'may', 'might', 'must', 'will']
for m in modals:
    print(m+':', fdist[m])
can: 94
could: 87
may: 93
might: 38
must: 53
will: 389
```

然后，选择 brown 语料库的不同部分，修改前面的例子，计数包含 wh 的词，如 what、when、where、who 和 why。

```
from nltk.corpus import brown
from nltk import FreqDist
news_text=brown.words(categories='news')
fdist=FreqDist([w.lower() for w in news_text]) #What 和 what
whs=['what', 'when', 'where', 'who', 'why']
for w in whs:
    print(w+':', fdist[w])
what: 95
when: 169
where: 59
who: 268
why: 14
```

下面，分组统计每一个感兴趣的文体，这里使用 NLTK 提供的带条件的频率分布函数。

```
>>> from nltk.corpus import brown
>>> from nltk import ConditionalFreqDist
>>> cfd=ConditionalFreqDist((genre, word)  #genre 文体，分组条件
...     for genre in brown.categories()  #统计所有分类(分类统计)
...     for word in brown.words(categories=genre)  #相关查询，统计所有单词的频率
... )
>>> genres=['news', 'religion', 'hobbies', 'science_fiction', 'romance', 'humor']
>>> #只看感兴趣的类和单词的统计情况
... modals=['can', 'could', 'may', 'might', 'must', 'will']
>>> cfd.tabulate(conditions=genres, samples=modals)
                  can  could  may  might  must  will
          news     93     86   66     38    50   389
      religion     82     59   78     12    54    71
       hobbies    268     58  131     22    83   264
science_fiction    16     49    4     12     8    16
       romance     74    193   11     51    45    43
         humor     16     30    8      8     9    13
```

新闻文体中最常见的情态动词是 will，而言情文体中最常见的情态动词是 could，其他文体中情态动词的特征不显著。反过来，最常见的情态动词是 will 的最有可能是新闻文体，最常见的情态动词是 could 的最有可能是言情文体。

2.1.4 多国语言语料库

NLTK 包含多国语言语料库,在使用这些语料库之前需要学习如何在 Python 中处理字符编码。udhr 语料库的 fileids 包括有关文件所使用的字符编码,如 UTF8 或者 Latin1,下面用条件频率分布来研究 udhr 语料库中英语和德语文本中的字长差异,如图 2-1 所示。

```
C:\Anaconda3>python
Python 3.4.1 |Anaconda 2.1.0 (64-bit)| (default, Sep 24 2014,
18:32:42) [MSC v.1600 64 bit (AMD64)] on win32
Type "help", "copyright", "credits" or "license" for more information.
>>> from nltk.corpus import udhr #udhr.zip解压
>>> from nltk import ConditionalFreqDist
>>> languages=['English', 'German_Deutsch']
>>> cfd=ConditionalFreqDist((lang, len(word)) for lang in languages for word in udhr.words(lang+'-Latin1'))
>>> cfd.plot(cumulative=True)
>>> print("%d%%" % (sum(cfd['English'].freq(l) for l in range(6))*100))
65%
>>> print("%d%%" % (sum(cfd['German_Deutsch'].freq(l) for l in range(6))*100))
58%
```

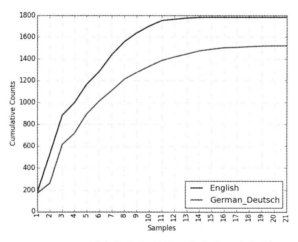

图 2-1 udhr 语料库中英语和德语文本中的字长差异

由图 2-1 可知,5 个或 5 个以下字母组成的词在英语文本中约占 65%,在德语文本中约占 58%。

```
>>> cfd
<ConditionalFreqDist with 2 conditions>
```

```
>>> cfd.conditions()
['German_Deutsch', 'English']
>>> cfd['English']
FreqDist({3: 358, 2: 340, 1: 185, 5: 169, 7: 157, 8: 118, 6: 117, 4: 114, 9: 80, 10: 63, ...})
>>> [(lang, len(word)) for lang in languages for word in udhr.words(lang+'-Latin1')]
[('English', 9), ('English', 11), ('English', 9), ('German_Deutsch', 3), ('German_Deutsch', 10), ('German_Deutsch', 9)]
'English':
9:2
11:1
'German_Deutsch':
3:1
9:1
10:1
```

2.1.5 文本语料库的结构

NLTK 中定义的基本语料库函数如表 2-2 所示,使用 help(nltk.corpus.reader)可以找到更多的文档。

表 2-2 NLTK 中定义的基本语料库函数

示例	描述
fileids()	语料库中的文件
categories()	语料库中的分类
raw(fileids=[f1,f2,f3])	指定文件的原始内容
words(fileids=[f1,f2,f3])	指定文件中的词汇
sents(fileids=[f1,f2,f3])	指定文件中的句子
abspath(fileid)	指定文件在磁盘上的位置
encoding(fileid)	文件的编码
open(fileid)	打开指定语料库文件的文件流
root()	到本地安装的语料库根目录的路径

各种语料库的访问方法如下:

```
from nltk.corpus import gutenberg
raw=gutenberg.raw("burgess-busterbrown.txt")
words=gutenberg.words("burgess-busterbrown.txt")
sents=gutenberg.sents("burgess-busterbrown.txt")
```

2.2 条件频率分布

语料文本分为几类时,我们可以计算每个类别独立的频率分布(分组统计),这样便可以研究类别之间的系统性差异。在 NLTK 中是用 ConditionalFreqDist 数据类型实现的,条件频率分布是频率分布的集合,每个频率分布有一个不同的条件,这个条件通常是文本的类别。

2.2.1 分组条件

频率分布计数文本中出现的词,条件频率分布需要给每个词关联一个条件,所以不是处理一个词序列,处理的是一个配对序列,每对的形式是(分组条件,单词)。

```
text1=['The', 'Fulton', 'County']
pairs1=[('news', 'The'), ('news', 'Fulton'), ('news', 'County')]
text2=['afraid', 'not']
pairs2=[('romance', 'afraid'), ('romance', 'not')]
```

2.2.2 按文体计数词汇

以 brown 语料库每一部分为条件,并按照每个条件计数词汇(分组统计)。FreqDist()以一个简单的列表作为输入,ConditionalFreqDist()以一个配对列表作为输入。

```
genre_word=[('news', 'The'), ('news', 'Fulton'), ('news', 'County'), ('romance', 'afraid'), ('romance', 'not')]
from nltk import ConditionalFreqDist
cfd=ConditionalFreqDist(genre_word)
cfd #<ConditionalFreqDist with 2 conditions>
cfd.conditions() #['news', 'romance']
cfd['news'] #FreqDist({'County': 1, 'Fulton': 1, 'The': 1})
cfd['romance']['afraid'] #1
```

拆开来看,只看 news 和 romance 两个文体,对于每个文体,我们遍历文体中的每个词以产生文体与词的配对。

```
from nltk.corpus import brown #解压 brown.zip
genre_word=[(genre, word) for genre in ['news', 'romance'] for word in brown.words(categories=genre)]
len(genre_word) #170576
```

因此,在下面的代码中我们可以看到,列表 genre_word 的前几个配对将是('news', word)的形式,而最后几个配对将是('romance',word)的形式。

```
genre_word[:3] #[('news', 'The'), ('news', 'Fulton'), ('news', 'County')]
#[('news', 'The'), ('news', 'Fulton'), ('news', 'County')]
genre_word[-3:] #[('romance', 'afraid'), ('romance', 'not'), ('romance', "'")]
#[('romance', 'not'), ('romance', "'"), ('romance', '.')]
```

现在，可以使用此配对列表创建一个 ConditionalFreqDist，并将它保存在一个变量 cfd 中。像往常一样，可以输入变量的名称来检查它，并确认它有两个条件。

```
from nltk import ConditionalFreqDist
cfd=ConditionalFreqDist(genre_word)
cfd #<ConditionalFreqDist with 2 conditions>
cfd.conditions() #['news', 'romance']
```

访问这两个条件，它们每一个都只是一个频率分布。

```
cfd['news'] #FreqDist({'the': 5580, ',': 5188, '.': 4030, 'of': 2849, ...})
cfd['romance'] #FreqDist({',': 3899, '.': 3736, 'the': 2758, ...})
list(cfd['romance'])[:4] #['blissful', 'MacIsaacs', 'mosaic', 'Pete']
cfd['romance']['could'] #193
```

2.2.3 绘制分布图和分布表

除了组合两个或两个以上的频率分布和更容易初始化之外，ConditionalFreqDist 还为制表和绘图提供了一些方法。

图 2-1 是基于下面的代码产生的一个条件频率分布绘制的，条件是语言的名称，图中的计数来源于词长，它利用了每一种语言的文件名是语言名称后面跟'-Latin1'(字符编码)。

```
>>> from nltk.corpus import udhr
>>> languages=['English', 'German_Deutsch']
>>> cfd=ConditionalFreqDist([(lang, len(word))
...  for lang in languages
...  for word in udhr.words(lang+'-Latin1')])
>>>   cfd.tabulate(conditions=['English',   'German_Deutsch'],
samples=range(10), cumulative=True)
                  0    1    2    3    4    5    6    7    8    9
        English   0  185  525  883  997 1166 1283 1440 1558 1638
 German_Deutsch   0  171  263  614  717  894 1013 1110 1213 1275
```

在 plot()和 tabulate()方法中，可以使用 conditions=parameter 来指定哪些条件显示。如果忽略它，所有条件都会显示。同样地，可以使用 samples=parameter 来限制要显示的样本。这使得载入大量数据到一个条件频率分布，然后通过选定条件和样品绘图或制表成为可能，这也使我们能全面控制条件和样本的显示顺序。例如，可以为两种语言和长度少于 10 个字符的词汇绘制累计频率数据表，上排最后一个单元格中数值的含义是英文文本中 9 个或少于 9 个字符长的词有 1 638 个。

处理 brown 语料库的 news 和 romance 文体，找出一周中最有新闻价值并且是最浪漫的日子。定义一个变量 days，包含星期几的列表，然后使用 cfd.tabulate(samples=days)为这些词的计数制表，接下来用绘图替代制表，可以在额外参数 conditions 的帮助下控制星期几输出的顺序。

```
>>> from nltk.corpus import brown
>>> cfd=ConditionalFreqDist([(genre, word)
... for genre in ['news', 'romance']
... for word in brown.words(categories=genre)])
>>> days=['Monday', 'Tuesday', 'Wednesday', 'Thursday', 'Friday', 'Saturday', 'Sunday']
>>> cfd.tabulate(samples=days)
         Monday Tuesday Wednesday Thursday Friday Saturday Sunday
    news     54      43        22       20     41       33     51
 romance      2       3         3        1      3        4      5
>>> from nltk.corpus import brown
>>> days=['Monday', 'Tuesday', 'Wednesday', 'Thursday', 'Friday', 'Saturday', 'Sunday']
>>> cfd=ConditionalFreqDist([(word, genre)
... for genre in ['news', 'romance']
... for word in brown.words(categories=genre)
... if word in days])
>>> cfd.tabulate(conditions=days)
           news romance
   Monday    54       2
  Tuesday    43       3
Wednesday    22       3
 Thursday    20       1
   Friday    41       3
 Saturday    33       4
   Sunday    51       5
```

2.2.4　使用双连词生成随机文本

我们可以使用条件频率分布创建一个双连词表，bigrams()函数接收一个词汇列表（上下文无关），并建立一个连续的词对列表（上下文有关）。在前文中，我们把每个词作为一个条件，对每个词创建它的后续词的频率分布（模型随机不随意），函数 generate_model()包含一个简单的循环来生成文本（依据模型）。调用这个函数时，选择一个词，如 living 作为初始内容（生成以 living 开头的文本，尽最大可能修复文本，呈现原文），进入循环后，输入变量 word 的当前值，重新设置 word 为上下文中最可能的标识符(word=cfdist[word].max())。下一次进入循环，我们使用哪个词作为新的初始内容呢？正如通过检查输出可以看到的一样，"of the land of the land" 中 of 后面最可能的词是 the，the 后面最可能的词是 land，land 后面最可能的词是 of（最佳情况是随机生成的文本和原文一样）。但是，这种简单的文本生成方法往往会在循环中被卡住，另一种方法是从可用的词汇中随机选择下一个词（从可能的词中）。

产生随机文本。此程序获得 genesis 文本中所有的双连词，然后构造一个条件频率分布来记录哪些词汇最有可能跟在给定词的后面，例如，living 后面最可能的词是 creature，creature 后面最可能的词是 that……像链表一样，用最有可能的方式建立联系，不是指针。generate_model()函数使用这些数据和种子词（模型）随机产生文本，这样产生的随机文本与原文最相似（随机不随意）。

```
C:\Anaconda3>python
Python 3.4.1 |Anaconda 2.1.0 (64-bit)| (default, Sep 24 2014,
18:32:42) [MSC v.1600 64 bit (AMD64)] on win32
Type "help", "copyright", "credits" or "license" for more information.
>>> from nltk import bigrams, ConditionalFreqDist
>>> from nltk.corpus import genesis
>>> def generate_model(cfd, word, num=15):  #文本长度 15
...     for i in range(num):
...         print(word, end=' ')
...         word=cfd[word].max()   #最可能的词，即频率最大者
...
>>> txt=genesis.words('english-kjv.txt')
>>> pairs=bigrams(txt)
>>> cfd=ConditionalFreqDist(pairs)
>>> print(cfd['living'])
<FreqDist with 6 samples and 16 outcomes>
>>> cfd['living']
FreqDist({'creature': 7, 'thing': 4, 'substance': 2, 'soul': 1, ',': 1, '.': 1})
>>> cfd['living'].max()
'creature'
```

living 后面可用的词汇有 creature，thing，substance，soul（解空间）等 6 个（6 samples，包括两个标点符号），它们出现在 living 后面的次数共 16 次（16 outcomes），最可能的词是 creature（最优解），它出现在 living 后面的次数最多，达 7 次，随机生成文本时，living 后接 creature 最合理。只可能从 creature，thing，substance，soul 中随机选择一个词接于 living 后面。

```
>>> generate_model(cfd, 'living')
living creature that he said , and the land of the land of the land
>>> from nltk import bigrams, ConditionalFreqDist
>>> from nltk.corpus import genesis
>>> from random import shuffle, choice
>>> def generate_model(cfd, word, num=15): #文本长度 15
...     for i in range(num):
...         print(word, end=' ')
...         #w=list(cfd[word])
...         #shuffle(w)
...         word=choice(list(cfd[word]))  #w[0]
#从可用的词汇中随机选择下一个词
>>> txt=genesis.words('english-kjv.txt')
>>> pairs=bigrams(txt)
>>> cfd=ConditionalFreqDist(pairs)
>>> generate_model(cfd, 'living')
living creature that Jacob sent over before thee as one another brother meeteth thee one
```

条件频率分布是一个对许多 NLP 任务都有用的数据结构，表 2-3 总结了它们常用的方法，包括定义、访问和可视化一个计数的条件频率分布的常用方法和习惯用法。

表 2-3 条件频率分布的常用方法和习惯用法

示例	描述
cfdist=ConditionalFreqDist(pairs)	从配对列表中创建条件频率分布
cfdist[condition]	此条件下的频率分布
cfdist[condition][sample]	此条件下给定样本的频率
cfdist.tabulate(samples，conditions)	指定样本和条件限制下制表
cfdist.plot(samples，conditions)	指定样本和条件限制下绘图

2.3 代码重用

2.3.1 使用文本编辑器创建程序

通常，使用文本编辑器组织多行程序，随后让 Python 一次性运行整个程序会更好。可以选择使用交互式解释器或文本编辑器来创建程序，使用解释器测试往往比较方便，修改一行代码直到达到期望效果。测试好之后，就可以将代码粘贴到文本编辑器，去除所有>>>和...提示符，继续扩展它，最后将程序保存为文件，这样以后就不用再重新输入了。给文件一个小而准确的名字，使用所有的小写字母，用下划线分割词汇，使用.py 文件名后缀。

2.3.2 函　数

一个函数是一个命名的代码块，用于执行一些明确的任务。一个函数通常被定义来使用一些称为参数的变量接受一些输入，并且它可能会产生一些结果，也称为返回值。下面使用关键字 def 加函数名以及所有输入参数来定义一个函数。函数的主体：

```
def lexical_diversity(text):
    return len(text)/len(set(text))
```

return 表示函数作为输出而产生的值。在这个例子中，函数所有的工作都在 return 语句中完成，下面是一个等价的定义，使用多行代码做同样的事，我们把参数名称从 text 变为 my_text_data。

```
def lexical_diversity(my_text_data):
    word_count=len(my_text_data)
    vocab_size=len(set(my_text_data))
    diversity_score=word_count/vocab_size
    return diversity_score
```

注意：已经在函数体内部创造了一些新的变量，这些是局部变量，不能在函数体外访问。现在已经定义一个名为 lexical_diversity 的函数，但只定义它不会产生任何输出，函数在调用之前不会做任何事情。

定义一个简单的函数来处理英文的复数词。函数 plural() 接受单数名词，产生一个复数形式（虽然它并不总是正确的）。这个函数试图生成任何英语名词的复数形式，关键字 def 后面跟着函数的名称，然后是包含在括号内的参数和一个冒号，函数体是缩进代码块。它试图识别词内的模式，相应地对词进行处理。例如，如果这个词以 y 结尾，删除它们并添加 ies。

```
>>> def plural(word): #定义
...     if word.endswith('y'):
...         return word[:-1]+'ies' #变 y 为 i, 加 es
...     elif word[-1] in 'sx' or word[-2:] in ['sh', 'ch']:
...         return word+'es'
...     elif word.endswith('an'):
...         return word[:-2]+'en'
...     else:
...         return word+'s'
```

```
...
>>> plural('family') #调用
'families'
>>> plural('boy')
'boies'
>>> plural('woman')
'women'
```

2.3.3 模　块

可以把劳动成果收集在一个单独的地方，在访问以前定义的函数时不必再复制。将函数保存到一个文件（如 textproc.py），随后便可以简单地以文件导入方式来访问该函数。

```
C:\Anaconda3\textproc.py:
def plural(word):
    if word.endswith('y'):
        return word[:-1]+'ies'
    elif word[-1] in 'sx' or word[-2:] in ['sh', 'ch']:
        return word+'es'
    elif word.endswith('an'):
        return word[:-2]+'en'
    else:
        return word+'s'


C:\Anaconda3>python
    Python 3.4.1 |Anaconda 2.1.0 (64-bit)| (default, Sep 24 2014, 18:32:42)
[MSC v.1600 64 bit (AMD64)] on win32
    Type "help", "copyright", "credits" or "license" for more information.
>>> from textproc import plural
>>> plural('fan') #fen
'fen'
```

显然，这里的复数函数明显存在错误，因为 fan 的复数是 fans。不用再重新输入这个函数的新版本，可以简单地编辑现有版本，因此，在任何时候复数函数只有一个版本，不会产生使用哪个版本的困扰。在一个文件中的变量和函数定义的集合被称为一个 Python 模块(module)，相关模块的集合称为一个包(package)。处理 brown 语料库的 NLTK 代码是一个模块，处理各种不同语料库代码的集合是一个包，NLTK 本身是包的集合，有时被称为一个库(library)。注意，创建一个包含一些 Python 代码的文件，一定不要将文件命名为 nltk.py，这可能会在导入时占据真正的 NLTK 包（当 Python 导入模块时，它会先查找当前目录）。

2.4 词典资源

词典资源是一个词和短语以及一些相关信息的集合，如词性和词义定义等相关信息。词典资源附属于文本，通常在文本的帮助下创建，例如，如果我们定义了一个文本 my_text，然后 vocab=sorted(set(my_text))建立 my_text 的词汇表，同时 word_freq=FreqDist(my_text)计数文本中每个词的频率，vocab 和 word_freq 都是简单的词汇资源。同样地，词汇索引为我们提供了有关词语用法的信息，可能在编写词典时有用。

2.4.1 词汇列表语料库

NLTK 包括一些仅仅包含词汇列表的语料库，我们可以用它来寻找文本语料中罕见或拼写错误的词汇。

过滤文本。此程序计算文本的词汇表，然后删除所有在现有的词汇列表中出现的元素，只留下罕见或拼写错误的词（UNK）。

```
>>> from nltk.corpus import words, nps_chat
>>> def unusual_words(txt):
...     txt_vocab=set(w.lower() for w in txt if w.isalpha())
...     english_vocab=set(w.lower() for w in words.words())
...     unusual=txt_vocab-english_vocab #集合的差运算
...     return sorted(unusual)
...
>>> unusual_words(nps_chat.words())
['aaaaaaaaaaaaaaaaa', 'aaahhhh', 'abortions', 'abou', ..., 'zzzzzzzz']
```

即时消息聊天会话语料库 nps_chat，用词随意，有很多罕见或拼写错误的词汇，如 aaaaaaaaaaaaaaaaa，aaahhhh 等。

还有一个停用词语料库，就是那些高频词汇，如 the，to，我们有时在进一步处理之前想要将它们从文档中过滤。停用词几乎没有什么词汇内容，而它们的出现会使区分文本变得更困难。

```
>>> from nltk.corpus import stopwords
>>> stopwords.words('english')
['i', 'me', 'my', 'myself', 'we', 'our', 'ours', ..., 'wouldn', "wouldn't"]
```

定义一个函数来计算文本中没有在停用词列表中的词的比例。

```
from nltk.corpus import stopwords, reuters
def content_fraction(txt):
```

```
    stopwords2=stopwords.words('english')
    content=[w for w in txt if w.lower() not in stopwords2]
    return len(content)/len(txt)

content_fraction(reuters.words())  #0.735240435097661
```

因此，在停用词的帮助下，我们筛选掉文本中 1/3 的词。注意，这里结合了两种不同类型的语料库，使用词典资源来过滤文本语料的内容。

2.4.2 发音词典

NLTK 中包括美国英语的 CMU 发音词典，它是为语音合成器使用而设计的。

```
>>> from nltk.corpus import cmudict
>>> entries=cmudict.entries()
>>> for entry in entries[42372:42374]:
...     print(entry)
...
('fire', ['F', 'AY1', 'ER0'])
('fire', ['F', 'AY1', 'R'])
>>> entries[0]
('a', ['AH0'])  #A 1 AH0
```

对每一个词，发音词典资源提供语音的代码，不同声音不同的标签，叫作音素。fire 有两个发音，单音节 F AY1 R 和双音节 F AY1 ER0。每个条目由两部分组成，我们可以用一个复杂的 for 语句来一个一个地处理这些。我们没有写 for entry in entries，而是用两个变量名 word 和 pron 替换 entry，现在，每次通过循环时，word 被分配条目的第一部分，pron 被分配条目的第二部分。

```
for word, pron in entries:
    if len(pron)==3:
        ph1, ph2, ph3=pron #取列表pron元素的方法
        if ph1=='P' and ph3=='T':
            print(word, ph2)
```

上面的程序扫描词典中那些发音包含三个音素的条目，如果条件为真，就将 pron 的内容分配给三个新的变量 ph1，ph2 和 ph3。注意：实现这个功能的语句形式并不多见。

这里是同样的 for 语句的另一个例子，这次使用内部的列表推导，这段程序找到所有发音结尾与 nicks 相似的词汇。可以使用此方法来找到押韵的词。

```
>>> syllable=['N', 'IH1', 'K', 'S'] #NICKS 1 N IH1 K S
>>> [word for word, pron in entries if pron[-4:]==syllable]
['knicks', "knicks'", "nick's", 'nicks', "nikk's", 'nix', 'nyx']
```

注意，有几种方法来拼读一个读音，nics，niks，nix 甚至 ntic's 加一个无声的 t，如词 atlantic's。

2.4.3 对照词表

表格词典的另一个例子是对照词表，NLTK 中包含 swadesh 核心词列表，语言标识符使用 ISO639 双字母码，如英语 en。

```
from nltk.corpus import swadesh
swadesh.fileids()
swadesh.words('en')
```

可以通过在 entries()方法中指定一个语言列表来访问多语言中的同源词，更进一步，可以把它转换成一个简单的词典。

```
fr2en=swadesh.entries(['fr', 'en'])
translate=dict(fr2en)
translate=nltk.Index(fr2en)
translate['chien']
translate['jeter']
```

通过添加其他源语言，可以让这个简单的翻译器更为有用。下面使用 dict()函数把德语-英语和西班牙语-英语对相互转换成一个词典，然后用这些添加的映射更新原来的翻译词典。

```
de2en = swadesh.entries(['de', 'en']) #German-English
es2en = swadesh.entries(['es', 'en']) #Spanish-English
translate.update(dict(de2en))
translate.update(dict(es2en))
translate['Hund'] #'dog'
translate['perro'] #'dog'
```

可以比较日耳曼语族和拉丁语族的不同。

```
languages = ['en', 'de', 'nl', 'es', 'fr', 'pt']
for i in [139, 140, 141, 142]:
    print(swadesh.entries(languages)[i])
```

小 结

（1）文本语料库是一个大型结构化文本的集合，NLTK 包含许多语料库，如 nltk.corpus. brown。

（2）有些文本语料库是分类的，例如，通过文体或者主题分类，有时候语料库的分类会相互重叠。

（3）条件频率分布是一个频率分布的集合，每个分布都有一个不同的条件，它们可以用于通过给定内容或者文体对词的频率计数。

（4）行数较多的 Python 程序应该使用文本编辑器来输入，保存为.py 后缀的文件，并使用 import 语句或者 python 命令来访问。

（5）WordNet 是一个面向语义的英语词典，由同义词的集合或称为同义词集(synsets)组成，并且组织成一个网络。

练 习

1. 定义一个列表变量 phrase，包含一个词的列表，验证本章描述的操作，包括加法、乘法、索引、切片和排序。

```
>>> phrase=['She', 'sells', 'sea', 'shells', 'by', 'the', 'sea', 'shore', '.']
>>> phrase+phrase #重载+运算符号
['She', 'sells', 'sea', 'shells', 'by', 'the', 'sea', 'shore', '.', 'She', 'sells', 'sea', 'shells', 'by', 'the', 'sea', 'shore', '.']
>>> phrase.__add__(phrase) #等效
['She', 'sells', 'sea', 'shells', 'by', 'the', 'sea', 'shore', '.', 'She', 'sells', 'sea', 'shells', 'by', 'the', 'sea', 'shore', '.']
>>> phrase*2
['She', 'sells', 'sea', 'shells', 'by', 'the', 'sea', 'shore', '.', 'She', 'sells', 'sea', 'shells', 'by', 'the', 'sea', 'shore', '.']
>>> phrase[0]
'She'
>>> phrase.index('She')
0
>>> phrase[0:]
['She', 'sells', 'sea', 'shells', 'by', 'the', 'sea', 'shore', '.']
>>> phrase
['She', 'sells', 'sea', 'shells', 'by', 'the', 'sea', 'shore', '.']
>>> sorted(phrase) #显示排序
['.', 'She', 'by', 'sea', 'sea', 'sells', 'shells', 'shore', 'the']
>>> phrase
['She', 'sells', 'sea', 'shells', 'by', 'the', 'sea', 'shore', '.']
>>> phrase.sort() #存储排序
```

```
>>> phrase
['.', 'She', 'by', 'sea', 'sea', 'sells', 'shells', 'shore', 'the']
```

2. 使用语料库模块处理 austen-persuasion.txt，计算出这本书中有多少词标识符？多少词类型？

```
from nltk.corpus import gutenberg
words=gutenberg.words('austen-persuasion.txt')
num_words=len(words)
num_words #98171
num_vocab=len(set(w.lower() for w in words))
num_vocab #5835
from nltk import word_tenize
raw=open('c:\\nltk_data\\corpora\\gutenberg\\austen-persuasion.txt').read()
words=word_tenize(raw)
num_words=len(words)
num_words #97918
num_vocab=len(set([word.lower() for word in words]))
num_vocab #5943
```

3. 使用 brown 语料库阅读器 nltk.corpus.brown.words()或网络文本语料库阅读器 nltk.corpus.webtext.words()来访问两个不同文体的一些样例文本。

```
nltk.corpus.brown.words()
nltk.corpus.webtext.words()
```

4. 使用 state_union 语料库阅读器，访问《国情咨文报告》的文本。计数每个文档中出现的 men、women 和 people。随着时间的推移，这些词的用法有什么变化？

```
from nltk.corpus import state_union
state_union.fileids()
text=nltk.Text(state_union.words('1945-Truman.txt'))
import nltk
text.dispersion_plot(['men', 'women', 'people'])
import nltk
from nltk.corpus import state_union
cfd=nltk.ConditionalFreqDist(
(target, fileid[:4])
for fileid in state_union.fileids()[:12]
for word in state_union.words(fileid)
for target in ['men', 'women', 'people']
```

```
      if target == word.lower()
)
#cfd.tabulate()
cfd.plot()
        1945 1946 1947 1948 1949 1950 1951 1953 1954 1955 1956 1957
    men    2   12    7    5    2    6    8    3    2    4    2    5
  people  10   49   12   22   15   15   10   17   15   26   30   11
  women    2    7    2    1    1    2    2    0    0    0    2    2

from nltk.corpus import brown as bn
from nltk.corpus import state_union as su
cfd=nltk.ConditionalFreqDist(
(target, fileid[:4])
for fileid in su.fileids()
for w in su.words(fileid)
for target in ['men', 'women', 'people'] if w.lower().startswith(target))

cfd.plot()
```

5. 考查一些名词的整体部分关系。注意，有 3 种整体部分关系，所以需要使用 member_meronyms(),part_meronyms(),substance_meronyms(),member_holonyms(),part_holonyms() 以及 substance_holonyms()。

```
from nltk.corpus import wordnet as wn
wn.synsets('smile')
#[Synset('smile.n.01'), Synset('smile.v.01'), Synset('smile.v.02')]
smile=wn.synset('smile.n.01')
smile
#Synet('smile.n.01')
smile.member_meronyms()
wn.synset('smile.n.01').definition
#'a facial expression characterized by turning up the corners of
the mouth; usually shows pleasure or amusement'
smile.hyponyms()
#[Synset('smirk.n.01'), Synset('simper.n.01')]
smile.hypernyms()
#[Synset('facial_expression.n.01')]
smile.part_meronyms()
smile.substance_meronyms()
smile.member_holonyms()
```

6. 在比较词表的讨论中，我们创建了一个对象叫作 translate，通过它你可以使用德语和意大利语词汇查找对应的英语词汇。这种方法可能会出现什么问题？你能提出一个办法来避免这个问题吗？

书上的做法是通过 entries()方法来指定一个语言链表来访问多种语言中的同源词，再把它转换成一个简单的词典。代码如下：

```
from nltk.corpus import swadesh
swadesh.fileids()
it2en=swadesh.entries(['it', 'en'])
de2en=swadesh.entries(['de', 'en'])
translate=dict(it2en)
translate.update(dict(de2en))
translate['Hund']
```

然而这个方法存在问题，原语言链表中有多对多关系的词，如 it2en 中的：

```
(u'tu, Lei', u'you(singular), thou')
(u'lui, egli', u'he')
(u'loro, essi', u'they')
(u'qui, qua', u'here')
(u'udire, sentire', u'hear')
(u'odorare, annusare', u'smell')
(u'dividere, separare', u'split')
(u'aguzzo, affilato', u'sharp')
(u'asciutto, secco', u'dry')
```

当输入 translate['tu']时并不会正确显示 you(singular)，thou，而是会报错 Traceback (most recent call last)。

```
  File "<stdin>", line 1, in <module>
KeyError: 'tu'
translate['tu, Lei']
```

解决思路：
遍历语言链表，当检测到有多对多关系时，将该元素进行处理后再加入原语言链表。
代码如下：

```
from nltk.corpus import swadesh
swadesh.fileids()
it2en=swadesh.entries(['it', 'en'])
de2en=swadesh.entries(['de', 'en'])
for key in it2en:
    if ', ' in key[0]:
```

```
            words=key[0].split(', ')
            for eachWord in words:
                newWord=(eachWord, key[1])
                it2en.append(newWord)

    for key in de2en:
        if ', ' in key[0]:
            words=key[0].split(', ')
            for eachWord in words:
                newWord=(eachWord, key[1])
                de2en.append(newWord)

    translate=dict(it2en)
    translate.update(dict(de2en))
    translate['tu']
    translate['tu, Lei']
```

7. 根据 Strunk 和 White 的 *Elements of Style*（由斯特伦克和怀特合著的《文体指南》），词 however 在句子开头使用是 "in whatever way" 或 "to whatever extent" 的意思，而没有 "nevertheless" 的意思。他们给出了正确用法的例子：However you advise him, he will probably do as he thinks best。使用词汇索引工具在各种文本中研究这个词的实际用法。

8. 在名字语料库上定义一个条件频率分布，显示哪个首字母在男性名字中比在女性名字中更常用。

```
cfd=nltk.ConditionalFreqDist(
    (fileid, name[0])
    for fileid in names.fileids()
    for name in names.words(fileid))
cfd.plot()
```

9. 挑选两个文本（text1 和 text2），研究它们在词汇、词汇丰富性、文体等方面的差异。你能找出几个在这两个文本中词意不同的词吗？例如，在《白鲸记》text1 与《理智与情感》text2 中的 monstrous。

```
import nltk
from nltk.bo import *
#text1=Text(gutenberg.words('melville-moby_dick.txt'))
#text2=Text(gutenberg.words('austen-sense.txt'))
```

关于词长的进一步分析可能帮助我们了解作者、文体或语言之间的差异。让我们来比较不同文体中的情态动词的用法。

10. 调查模式分布表，寻找其他模式。试着用你自己对不同文体的印象理解来解释它们。你能找到其他封闭的词汇归类，展现不同文体的显著差异吗？

封闭类（CLOSED CLASSES）

介词：of，at，in，without，in spite of

代词：he，anybody，they，one，which

限定词：the，a，that，every，some

连词：and，that，when，although

情态动词：can，must，will，could

基本动词：be，have，do

11. CMU 发音词典包含某些词的多个发音。它包含多少种不同的词？具有多个可能的发音的词在这个词典中的比例是多少？

```
import nltk
entries=nltk.corpus.cmudict.entries()
len(entries) #133737
fd=nltk.FreqDist([entry[0] for entry in entries])
len([w for w, f in fd.items() if f>1]) #9241
100.0*9241/133737
import nltk
from nltk.corpus import cmudict
entries=cmudict.entries()
raw=cmudict.words()
print "the words number is %d, the unique words number is %d" % (len(entries), len(set(raw)))
dictionary=[]
for word in entries:
    dictionary.append(word[0])

print len(dictionary)
print len(set(dictionary))
import nltk
from __future__ import division

entries = nltk.corpus.cmudict.entries()
ky = entries[0][0]
cnt = 0
for key, pron in entries:    # 不同发音计数
    if ky != key:
        ky = key
```

```
        cnt += 1

length = len(entries)    # CMU 字典总数
lenSets = [len(pron) for key, pron in entries]    # 多种发音的计数
percentage = []

for i in range(1, max(lenSets) + 1):
    percentage.append(lenSets.count(i) / length * 100)

print cnt
for i in percentage:
    print '%.3f %%' % i    # 格式化输出
from collections import Counter
Counter(fd).most_common()[0][1] #???
```

12. 没有下位词的名词同义词集所占的百分比是多少？你可以使用 wn.all_synsets('n')得到所有名词同义词集。

```
import nltk
from nltk.corpus import wordnet as wn
from __future__ import division
all_noun_dict=wn.all_synsets('n')
all_noun_num=len(set(all_noun_dict)) #82115
noun_without_hypon=filter(lambda    ss:len(ss.hyponyms())<=0,
wn.all_synsets('n'))
noun_without_num=len(list(noun_without_hypon)) #65422
print 'There are %d nouns, and %d nouns without hyponyms, the
percentage is %f' % (all_noun_num, noun_without_num, noun_without_num/
all_noun_num*100)
#There are 82115 nouns, and 65422 nouns without hyponyms, the
percentage is 79.671193
nltk.app.wordnet()
turnout
```

13. 定义函数 supergloss(s)，使用一个同义词集 s 作为它的参数，返回一个字符串，包含 s 的定义和 s 所有的上位词与下位词的定义的连接字符串。

```
from nltk.corpus import wordnet as wn
#nltk 中也有一个 wordnet!
def supergloss(s):
    d=s.definition
```

```
    h1=s.hypernyms()
    h2=s.hyponyms()
    r=d
    for h in h1+h2:
        r+=h.definition
    return r

s=wn.synset('car.n.01')
supergloss(s)
```

14. 写一个程序，找出所有在 brown 语料库中至少出现 3 次的词。

```
import nltk
from nltk.corpus import brown
words=brown.words()
fd=nltk.FreqDist(words)
[w for w, f in fd.items() if f>=3]
```

15. 写一个程序，生成如表 1-1 所示的词汇多样性得分表（例如，标识符数除以类型数，我们定义为单位类型标识符数，作为度量标准词汇多样性的一个指标，类似于商品的单价）。brown 语料库文体的全集(nltk.corpus.brown.categories())中，哪个文体词汇多样性最低（单位类型的标识符数最多）？这是你所期望的吗？

```
import nltk
from nltk.corpus import brown
from __future__ import division
def word_diversity(words):
    return len(words)/len(set(words))

for category in brown.categories():
    diversity=word_diversity(brown.words(categories=category))
    print '%10s\t%.2f' % (category, diversity)
 adventure          7.81
belles_lettres      9.40
 editorial          6.23
   fiction          7.36
government          8.57
   hobbies          6.90
     humor          4.32
```

```
       learned           10.79
          lore            7.61
       mystery            8.19
          news            6.99
       religion           6.18
        reviews           4.72
        romance           8.28
science_fiction           4.48
```

learned 词汇多样性最低（10.79），语言严密准确；humor 词汇多样性最高（4.32），语言丰富。符合期望。

16. 写一个函数，找出一个文本中最常出现的 50 个词（停用词除外）。

```
import nltk
from nltk.bo import *
from nltk.corpus import stopwords
from collections import Counter
def fun(text):
    fd=nltk.FreqDist([w.lower() for w in text if w not in stopwords.words('english')])
    return [w for w, _ in Counter(fd).most_common(50)]

fun(text1)
```

17. 写一个程序，输出一个文本中 50 个最常见的双连词（相邻词对），忽略包含停用词的双连词。

```
import nltk
from nltk.bo import *
from nltk.corpus import stopwords
from collections import Counter
def fun(text):
    fd=nltk.FreqDist([(w1, w2) for w1, w2 in nltk.bigrams(text) if w1 not in stopwords.words('english') and w2 not in stopwords.words('english')])
    return [w for w, _ in Counter(fd).most_common(50)]

text='Excuse me . Is there any summer resort around the town ?'
text=nltk.word_tenize(text)
fun(text)
```

18. 写一个程序，按文体创建一个词频表，以 2.1 节给出的词频表为范例，选择你自己的词汇，尝试找出那些在一个文体中很突出或很缺乏的词汇，并讨论研究结果。

```
from nltk.corpus import brown
from nltk import ConditionalFreqDist
from nltk.corpus import stopwords
stopwords_set=set(stopwords.words(fileids=u'english'))
pairs=[(wenti, word.lower()) for wenti in brown.categories() for word in brown.words(categories=wenti) if word.lower() not in stopwords_set]
cfd=ConditionalFreqDist(pairs)
for wenti in brown.categories():
    fdist=cfd[wenti]
    sorted_words=sorted(fdist.keys(), key=lambda x:fdist[x], reverse=True)
    print wenti
    print '===>',
    print ', '.join(sorted_words[:10])
    print '===>',
    print ', '.join(sorted_words[-10::])
```

高频次不具备分类性，应考虑类间分布度量因子

19. 写一个函数 word_freq()，用一个词和 brown 语料库中的一个部分的名字作为参数，计算这部分语料中词的频率。

```
import nltk
from nltk.corpus import brown
def word_freq(word, name):
    fd=nltk.FreqDist(brown.words(categories=name))
    return fd.freq(word)
def freq(word, category):
    text=nltk.Text(brown.words(categories=category))
    return 1.0*text.count(word)/len(text)
```

20. 写一个程序，估算一个文本中的音节数，利用 CMU 发音词典。

```
from nltk.corpus import cmudict
from nltk.bo import *
def musiccounts(text):
    entries=nltk.corpus.cmudict.entries()
    prons=[pron for w in text for w, pron in entries]
    return len(prons)
```

21. 定义一个函数 hedge(text)，处理一个文本和产生一个新的版本在每三个词之间插入一个词 like。

```
def hedge(sent):
    new_sent=[]
    for insert_index in range(3, len(sent), 3):
        new_sent.extend(sent[insert_index-3:insert_index]+['like'])
    new_sent.extend(sent[insert_index:])
    return new_sent

sent=['The', 'family', 'of', 'Dashwood', 'had', 'long', 'been', 'settled', 'in', 'Sussex', '.']
hedge(sent)
```

22. 齐夫定律：f(w)是一个自由文本中的词 w 的频率。假设一个文本中的所有词都按照它们的频率排名，频率最高的在最前面。齐夫定律指出一个词类型的频率与它的排名成反比（即 f×r=k，k 是某个常数）。例如，最常见的第 50 个词类型出现的频率应该是最常见的第 150 个词型出现频率的 3 倍。（r1=50，r2=150，50f1=150f2，f1=3f2）

（1）写一个函数来处理一个大文本，使用 pylab.plot 画出相对于词排名的词的频率。你认可齐夫定律吗？所绘的线的极端情况是怎样的？（提示：使用对数刻度会有帮助）

```
from nltk.corpus import inaugural
from nltk import FreqDist
from math import log10
from pylab import *
r=[float('%.2f' % log10(r)) for r in range(1, 151)]
words=inaugural.words()
fd=FreqDist([w.lower() for w in words])
fs=fd.values()[:150]
f=[float('%.2f' % log10(f)) for f in fs]
xlabel('r')
ylabel('f')
axis([min(r), max(r), min(f), max(f)])
title(u'f×r=k')
plot(r, f)
show()
```

（2）随机生成文本，如使用 random.choice("abcdefg")，注意要包括空格字符，需要事先使用 import random。使用字符串连接操作将字符累积成一个很长的字符串，然后为这个字符串分词，产生前面的齐夫图，比较这两个图。

```
from random import *
from nltk import *
from math import log10
from pylab import *
text=""
for i in range(600000):
    text+=choice("abcdefg ")

words=word_tenize(text)
r=[float('%.2f' % log10(r)) for r in range(1, 151)]
fd=FreqDist([w.lower() for w in words])
fs=fd.values()[:150]
f=[float('%.2f' % log10(f)) for f in fs]
xlabel('r')
ylabel('f')
axis([min(r), max(r), min(f), max(f)])
title(u'f×r=k')
plot(r, f)
show()
```

实验结果：

第一问实验认可齐夫定律。

第二问实验不符合齐夫定律。

23. 修改文本生成程序，进一步完成下列任务。

（1）在一个词列表中存储 n 个最相似的词（就是 cfdist[word]），使用 random.choice()从列表中随机选取一个词。（需要事先使用 import random）

```
import nltk, random
def generate_model(cfdist, word, num=15):
    for i in range(num):
        print word,
        words=cfdist[word].keys()[:4]
        word=random.choice(words)  #一定程度解决了循环卡住问题

text=nltk.corpus.genesis.words('english-kjv.txt')
bigrams=nltk.bigrams(text)
cfd=nltk.ConditionalFreqDist(bigrams)  #训练模型
generate_model(cfd, 'living')
```

（2）选择特定的文体，如 brown 语料库中的一部分、《创世纪》翻译、古腾堡语料库中的文本或一个网络文本。在此语料上训练一个模型（实际就是 cfd），产生随机文本。你可能要尝试不同的起始字。文本的可理解性如何？讨论这种方法产生随机文本的长处和短处。

```
import nltk, random
def generate_model(cfdist, word, num=150):
    for i in range(num):
        print word,
        words=cfdist[word].keys()[:4]
        word=random.choice(words)  #一定程度解决了循环卡住问题

text=nltk.corpus.brown.words(categories='news')
bigrams=nltk.bigrams(text)
cfd=nltk.ConditionalFreqDist(bigrams)
voc=list(set([w for w in text]))
word=random.choice(voc)
generate_model(cfd, word)
```

（3）现在使用两种不同文体训练系统，使用混合文体文本做实验，并讨论观察结果。

```
import nltk
def generate_model(cfdist, word, num=150):
    for i in range(num):
        print word,
        words=cfdist[word].keys()[:4]
        word=random.choice(words)  #一定程度解决了循环卡住问题

text1=nltk.corpus.brown.words(categories='news')
text2=nltk.corpus.brown.words(categories='romance')
text=text1+text2 #???
bigrams=nltk.bigrams(text)
cfd=nltk.ConditionalFreqDist(bigrams)
voc=list(set([w for w in text]))
word=random.choice(voc)
generate_model(cfd, word)
```

24. 定义一个函数 find_language()，用一个字符串作为其参数，返回包含这个字符串作为词汇的语言的列表。使用《世界人权宣言》(udhr) 的语料，将搜索限制在 Latin-1 编码的文件中。

```
import nltk
from nltk.corpus import udhr
```

```
def find_language(w):
    ids=udhr.fileids()
    ids=[id for id in ids if id.endswith('Latin1')]
    l=[id[:-7] for id in ids if w in set(w.lower() for w in udhr.words(id))]
    return l

find_language('the')
['English', 'Paez', 'Rukonzo_Konjo', 'Yao']
```

25. 名词上位词层次的分枝因素是什么？也就是说，对于每一个具有下位词——上位词层次中的子女的名词同义词集，它们平均有几个下位词？可以使用 wn.all_synsets('n')获得所有名词同义词集。

```
import nltk
from nltk.corpus import wordnet as wn
from __future__ import division
ss=filter(lambda x : len(x.hyponyms())>0, wn.all_synsets('n'))
n=len(list(ss)) #16693
m=sum(map(lambda x : len(x.hyponyms()), ss)) #75850
m/n #4.543820763194153
```

26. 一个词的多义性是指它所有含义的个数。利用 WordNet，使用 len(wn.synsets('dog', 'n'))可以判断名词 dog 有 7 种含义。试计算 WordNet 中名词、动词、形容词和副词的平均多义性。

```
from nltk.corpus import wordnet as wn
len(wn.synsets('dog', 'n')) #7
from nltk.corpus import wordnet as wn
types=[('n', '名词'), ('v', '动词'), ('a', '形容词'), ('r', '副词')]
for type in types:
    synsets=wn.all_synsets(type[0]) #收集所有指定类型的同义词集
    lemmas=[] #
    for synset in synsets: #遍历同义词集
        for lemma in synset.lemmas:
            lemmas.append(lemma.name) #收集同义词集内的词条
    lemmas=set(lemmas) #词条去重
    print type[1], '的词条有', len(lemmas), '条' #打印词条总数
    count=0 #
    for lemma in lemmas: #遍历词条
```

```
            count+=len(wn.synsets(lemma, type[0]))  #得到每个词条的同类型
所含意义的个数，求和
    print '词条的意义总数为:', count #打印词集内该类型词条的意义总数
    print type[1], '的平均多义性为:', count/len(lemmas) #打印该类型的
平均多义性
```

27. 使用预定义的相似性度量之一给下面的每个词对的相似性打分。按相似性减少的顺序排名。你的排名与这里给出的顺序有多接近？（Miller & Charles，1998）实验得出的顺序：

```
car-automobile, gem-jewel, journey-voyage, boy-lad, coast-shore,
asylum-madhouse, magician-wizard, midday-noon, furnace-stove, food-
fruit, bird-cock, bird-crane, tool-implement, brother-monk, lad-brother,
crane-implement, journey-car, monk-oracle, cemetery-woodland, food-
rooster, coast-hill, forest-graveyard, shore-woodland, monk-slave,
coast-forest, lad-wizard, chord-smile, glass-magician, rooster-voyage,
noon-string.
    import nltk
    from nltk.corpus import wordnet as wn
    wn.synsets('car') #[Synset('car.n.01'), Synset('car.n.02'), Synset
('car.n.03'), Synset('car.n.04'), Synset('cable_car.n.01')]
    wn.synsets('automobile') #[Synset('car.n.01'), Synset('automobile.v.01')]
    car=wn.synset('car.n.01')
    automobile=wn.synset('car.n.01')
    car.path_similarity(automobile) #1.0
```

阅读材料

1. Python 基于 WordNet 实现词语相似度计算分析

WordNet 是由普林斯顿大学的心理学家、语言学家和计算机工程师联合设计的一种基于认知语言学的英语词典。它不仅是把单词以字母顺序排列，而且按照单词的意义组成一个"单词的网络"。它是一个覆盖范围广的英语词汇义网。名词、动词、形容词和副词各自组织成一个同义词的网络，每个同义词集合都代表一个基本的语义概念，并且这些集合之间也由各种关系连接。WordNet 包含描述概念含义、一义多词、一词多义、类别归属、近义、反义等问题。鉴于 WordNet 本身的性质，我们想到了可以借助于同一词典网络的形式来计算词语之间的相似度，具体的实现很简单，核心的思想就是同义词汇数据。具体的代码实现如下：

```
Python 3.7.3 (default, Mar 27 2019, 17:13:21) [MSC v.1915 64 bit (AMD64)] :: Ana
conda, Inc. on win32
>>> import sys
>>> import numpy as np
>>> import pandas as pd
>>> from scipy import stats
>>> from nltk.corpus import wordnet as wn
>>> from sklearn.preprocessing import MinMaxScaler, Imputer
Traceback (most recent call last):
  File "<stdin>", line 1, in <module>
ImportError: cannot import name 'Imputer' from 'sklearn.preprocessing' (C:\Users
\LYS\Anaconda3\lib\site-packages\sklearn\preprocessing\__init__
.py)

Python 2.7.11 |Anaconda 2.5.0 (64-bit)| (default, Jan 29 2016, 14:26:21) [MSC v.
1500 64 bit (AMD64)] on win32
Type "help", "copyright", "credits" or "license" for more information.
Anaconda is brought to you by Continuum Analytics.
Please check out: http://continuum.io/thanks and https://anaconda.org
>>> #encoding:utf-8
... from __future__ import division
>>> '''
... Anaconda2-2.5.0-Windows-x86_64
```

```
...    功能:基于WordNet的词语相似度计算分析
...    '''
'\nAnaconda2-2.5.0-Windows-x86_64\n\xb9\xa6\xc4\xdc\xa3\xba\xbb
\xf9\xd3\xdaWordN
    et\xb5\xc4\xb4\xca\xd3\xef\xcf\xe0\xcb\xc6\xb6\xc8\xbc\xc6\xcb\
\xe3\xb7\xd6\xce\x
f6\n'
>>> import sys
>>> import numpy as np
>>> import pandas as pd
>>> from scipy import stats
>>> from nltk.corpus import wordnet as wn
>>> from sklearn.preprocessing import MinMaxScaler, Imputer
>>> reload(sys)
<module 'sys' (built-in)>
>>> sys.setdefaultencoding('utf-8')

>>> data='data.csv'
>>> data
'data.csv'
>>> data=pd.read_csv(data)
>>> data
    word1    word2     similarity
0   apple    orange    1
1   iphone   phone     2
2   house    horse     1
3   car      bicycle   1
4   human    woman     3
5   big      huge      4
6   rain     wind      1
7   spider   crawl     1
8   fire     fireman   1
9   flood    blood     1


#encoding:utf-8
from __future__ import division
'''
Anaconda2-2.5.0-Windows-x86_64
功能:基于WordNet的词语相似度计算分析
```

```python
'''
import sys
import numpy as np
import pandas as pd
from scipy import stats
from nltk.corpus import wordnet as wn
from sklearn.preprocessing import MinMaxScaler, Imputer
reload(sys)
sys.setdefaultencoding('utf-8')

def loadData(data='data.csv'):
    '''
    加载数据集
    '''
    data=pd.read_csv(data)
    word_list=np.array(data.iloc[1:, [0, 1]])
    self_sim_res=np.array(data.iloc[1:, [2]])
    return word_list, self_sim_res

def calWordSimilarity(word_list, self_sim_res, res_path='wordnetResult.csv'):
    '''
    计算词语相似度
    '''
    self_sim_matrix=np.zeros((len(self_sim_res), 1))
    for i, word_pair in enumerate(word_list):
        word1, word2=word_pair
        count=0
        synsets1=wn.synsets(word1)
        synsets2=wn.synsets(word2)
        print 'synsets1: ', synsets1
        print 'synsets2: ', synsets2
        for synset1 in synsets1:
            for synset2 in synsets2:
                score=synset1.path_similarity(synset2)
                if score is not None:
                    self_sim_matrix[i, 0]+=score
                    count+=1
                else:
                    pass
        self_sim_matrix[i, 0]=self_sim_matrix[i, 0]*1.0/count
```

```
        imputer=Imputer(missing_values='NaN', strategy='mean', axis=0)
        imputer_list=imputer.fit_transform(self_sim_matrix)
        scaler=MinMaxScaler(feature_range=(0.0, 10.0))
        imputer_list_scale=scaler.fit_transform(imputer_list)
        (coefidence, p_value)=stats.spearmanr(self_sim_res, imputer_list_scale)
        print 'coefidence: ', coefidence
        print 'p_value: ', p_value
        submitData=np.hstack((word_list, self_sim_res, imputer_list_scale))
        (pd.DataFrame(submitData)).to_csv(res_path, index=False,
                header=["Word1", "Word2", "originalSim", "wordnetSim"])

if __name__=='__main__':
    word_list, self_sim_res=loadData(data='data.csv')
    calWordSimilarity(word_list, self_sim_res, res_path='wordnetResult.csv')
```

下面来看具体的计算应用实例。原始数据文件的格式如下:

word1	word2	similarity
apple	orange	1
iphone	phone	2
house	horse	1
car	bicycle	1
human	woman	3
big	huge	4
rain	wind	1
spider	crawl	1
fire	fireman	1
flood	blood	1

其中，前两列分别为需要计算的词汇，最后一列是人为给定的初始相似度数据。简单的测试结果输出如下:

```
synsets1:  []
synsets2:  [Synset('telephone.n.01'), Synset('phone.n.02'), Synset('earphone.n.01'), Synset('call.v.03')]
synsets1:   [Synset('house.n.01'), Synset('firm.n.01'), Synset
```

('house.n.03'), Synset('house.n.04'), Synset('house.n.05'), Synset('house.n.06'), Synset('house.n.07'), Synset('sign_of_the_zodiac.n.01'), Synset('house.n.09'), Synset('family.n.01'), Synset('theater.n.01'), Synset('house.n.12'), Synset('house.v.01'), Synset('house.v.02')]

synsets2: [Synset('horse.n.01'), Synset('horse.n.02'), Synset('cavalry.n.01'), Synset('sawhorse.n.01'), Synset('knight.n.02'), Synset('horse.v.01')]

synsets1: [Synset('car.n.01'), Synset('car.n.02'), Synset('car.n.03'), Synset('car.n.04'), Synset('cable_car.n.01')]

synsets2: [Synset('bicycle.n.01'), Synset('bicycle.v.01')]

synsets1: [Synset('homo.n.02'), Synset('human.a.01'), Synset('human.a.02'), Synset('human.a.03')]

synsets2: [Synset('woman.n.01'), Synset('woman.n.02'), Synset('charwoman.n.01'), Synset('womanhood.n.02')]

synsets1: [Synset('large.a.01'), Synset('big.s.02'), Synset('bad.s.02'), Synset('big.s.04'), Synset('big.s.05'), Synset('big.s.06'), Synset('boastful.s.01'), Synset('big.s.08'), Synset('adult.s.01'), Synset('big.s.10'), Synset('big.s.11'), Synset('big.s.12'), Synset('big.s.13'), Synset('big.r.01'), Synset('boastfully.r.01'), Synset('big.r.03'), Synset('big.r.04')]

synsets2: [Synset('huge.s.01')]

synsets1: [Synset('rain.n.01'), Synset('rain.n.02'), Synset('rain.n.03'), Synset('rain.v.01')]

synsets2: [Synset('wind.n.01'), Synset('wind.n.02'), Synset('wind.n.03'), Synset('wind.n.04'), Synset('tip.n.03'), Synset('wind_instrument.n.01'), Synset('fart.n.01'), Synset('wind.n.08'), Synset('weave.v.04'), Synset('wind.v.02'), Synset('wind.v.03'), Synset('scent.v.02'), Synset('wind.v.05'), Synset('wreathe.v.03'), Synset('hoist.v.01')]

synsets1: [Synset('spider.n.01'), Synset('spider.n.02'), Synset('spider.n.03')]

synsets2: [Synset('crawl.n.01'), Synset('crawl.n.02'), Synset('crawl.n.03'), Synset('crawl.v.01'), Synset('crawl.v.02'), Synset('crawl.v.03'), Synset('fawn.v.01'), Synset('crawl.v.05')]

synsets1: [Synset('fire.n.01'), Synset('fire.n.02'), Synset('fire.n.03'), Synset('fire.n.04'), Synset('fire.n.05'), Synset('ardor.n.03'),

```
Synset('fire.n.07'), Synset('fire.n.08'), Synset('fire.n.09'), Synset
('open_fire.v.01'), Synset('fire.v.02'), Synset('fire.v.03'), Synset
('displace.v.03'), Synset('fire.v.05'), Synset('fire.v.06'), Synset
('arouse.v.01'), Synset('burn.v.01'), Synset('fuel.v.02')]
    synsets2: [Synset('fireman.n.01'), Synset('stoker.n.02'), Synset
('reliever.n.03'), Synset('fireman.n.04')]
    synsets1: [Synset('flood.n.01'), Synset('flood.n.02'), Synset
('flood.n.03'), Synset('flood.n.04'), Synset('flood.n.05'), Synset
('flood_tide.n.02'), Synset('deluge.v.01'), Synset('flood.v.02'),
Synset('flood.v.03'), Synset('flood.v.04')]
    synsets2: [Synset('blood.n.01'), Synset('blood.n.02'), Synset('rake.
n.01'), Synset('lineage.n.01'), Synset('blood.n.05'), Synset('blood.
v.01')]
    coefidence:  0.0，置信度
    p_value:  1.0
```

计算结果数据如下：

Word1	Word2	originalSim	wordnetSim
iphone	phone	2	3.414788047
house	horse	1	2.423915996
car	bicycle	1	10
human	woman	3	1.699208352
big	huge	4	3.414788047
rain	wind	1	4.123821286
spider	crawl	1	0
fire	fireman	1	2.567593558
flood	blood	1	3.088977134

相比于原始数据文件，这里多了最后一列，是基于 WordNet 计算的结果。

2. WordNet 的安装与使用

WordNet 的安装方法：

（1）安装 NLTK：

```
pip install nltk
```

（2）用 NLTK 的 downloader 下载 "WordNet"，获取相关的数据。

```
import nltk
nltk.download('wordnet')  #这是英文版 WordNet
```

(3）如果要使用中文版 WordNet，需要再下载一个组件"omw"。

```
nltk.download('omw') #omw 代表 Open Multilingual Wordnet
[nltk_data] Downloading package omw to C:\nltk_data...
[nltk_data]   Package omw is already up-to-date!
True
```

WordNet 的使用方法：

（1）词义查询：

```
Python 2.7.11 |Anaconda 2.5.0 (64-bit)| (default, Jan 29 2016, 14:26:21) [MSC v.1500 64 bit (AMD64)] on win32
>>> from nltk.corpus import wordnet as wn
>>> #获得单个词的定义查询
... apple = wn.synset('apple.n.01')
>>> print(apple.definition())
fruit with red or yellow or green skin and sweet to tart crisp whitish flesh
>>> #获得该词的所有词性及解释下的定义
... word = 'apple'
>>> for w in wn.synsets(word):
...     print(w.definition())
...
fruit with red or yellow or green skin and sweet to tart crisp whitish flesh
native Eurasian tree widely cultivated in many varieties for its firm rounded edible fruits
>>> wn.synsets(word)
[Synset('apple.n.01'), Synset('apple.n.02')]
```

因为中文在查询时，本质上还是映射到英文语义上去，所以不能直接用类似"秘密.n.01"这种形式，只能用 synsets 来查，但 synset 没有'lang'参数。

```
>>> word = u'秘密'
>>> print('origin word:', word)
('origin word:', u'\u79d8\u5bc6')
>>> if len(wn.synsets(word, lang='cmn')) == 0:
...     print('No this word')
...
>>> for w in wn.synsets(word, lang='cmn'):
...     print(w)
...     print(w.definition())
```

```
...
Synset('mystery.n.01')
something that baffles understanding and cannot be explained
Synset('secret.n.01')
something that should remain hidden from others (especially information that is not to be passed on)
Synset('privacy.n.02')
the condition of being concealed or hidden
```

一个词可能同时具有动词、名词等多种词性，而且每个词性下可能具有多种解释。例如，在查询"privacy"一词时："privacy.n.01"代表"the quality of being secluded from the presence or view of others"，"privacy.n.02"代表"the condition of being cncealed or hidden"，n代表名词，v代表动词，数字代表第几个。

（2）同义词查询

```
from nltk.corpus import wordnet as wn
#方法一：
print(wn.synset('apple.n.01').lemma_names())
#方法二：
for w in wn.synsets('apple'):
    print(w.lemma_names())

word = '秘密'
for w in wn.synsets(word, lang='cmn'):
    print(w.lemma_names())
```

这里相当于把中文的"秘密"与英文中的词做了一个对应，对应到三个名词，分别是上面提到的'mystery.n.01'，'secret.n.01'和'privacy.n.02'。在找同义词时，分别找到了"秘密"这个中文词对应的三个英文词的同义词。

（3）其他查询：

```
hypernyms()  #上位(父类)
hyponyms()   #下位(子类)
lemma_names()  #同义
antonyms()   #反义
entailments()  #蕴含关系
part_meronyms()  #部分
substance_meronyms()  #实质
member_holonyms()  #成员
```

第 3 章 文本处理

3.1 从网络和本地访问文本

3.1.1 从网络访问文本

在网络上浏览 60 000 本免费在线书籍的目录，获得 ASCII 码文本文件的 URL，用如下方式进行访问。

```
#Python 3.4.1 |Anaconda 2.1.0 (64-bit)
>>> from urllib.request import urlopen
>>> url="https://www.gutenberg.org/files/2554/2554-0.txt" #注意网址的变化
>>> raw=urlopen(url).read() #有些版本出错
>>> type(raw)
<class 'bytes'>
>>> len(raw)
1201520
>>> raw[:76]
b'\xef\xbb\xbfThe Project Gutenberg eBook of Crime and Punishment, by Fyodor Dostoevsky'
>>> raw=raw.decode('utf8')
>>> type(raw)
<class 'str'>
>>> len(raw)
1176812
>>> raw[:76] #文件开头有一个BOM
'\ufeffThe Project Gutenberg eBook of Crime and Punishment, by Fyodor Dostoevsky\r\n'
```

read()需要几秒钟来下载这本书。变量 raw 包含一个有 1 176 812 个字符的字符串，使用 type(raw)可以看到它是一个字符串，这是这本书原始的内容，包含很多我们不感兴趣的细节，如空格、换行符和空行等。注意，文件中行尾的\r\n 是 Python 用来显示回车和换行字符的方式，这个文件是在 Windows 系统中创建的，在 Unix 系统中创建的文件，文件中行尾是\n。对于语言处理，要将字符串分解为词和标点符号，这一步被称为分词。

```
from nltk import word_tokenize
from urllib.request import urlopen
url="https://www.gutenberg.org/files/2554/2554-0.txt"
raw=urlopen(url).read()
raw=raw.decode('utf8')
tokens=word_tokenize(raw)
type(tokens) #<class 'list'>
len(tokens) #244780
tokens[1:10] #粒度是单词
```

注意，NLTK 需要分词，再进行下一步。从这个列表创建一个 NLTK 文本。

```
>>> from nltk import word_tokenize
>>> from urllib.request import urlopen
>>> url="https://www.gutenberg.org/files/2554/2554-0.txt"
>>> raw=urlopen(url).read()
>>> raw=raw.decode('utf8')
>>> tokens=word_tokenize(raw)
>>> type(tokens) #<class 'list'>
<class 'list'>
>>> len(tokens) #244780
244780
>>> tokens[1:10] #粒度是单词
['Project', 'Gutenberg', 'eBook', 'of', 'Crime', 'and', 'Punishment', ',', 'by']
```

注意：从 gutenberg 项目下载的每个文本都包含一个首部，里面有文本的名称、作者、扫描和校对文本的人的名字、许可证等信息，有时这些信息出现在文件末尾页脚处，不能可靠地检测出文本内容的开始和结束。因此，在从原始文本中挑出内容之前，需要手工检查文件，以发现标记内容开始的独特的字符串 PART I。

```
>>> raw.find("PART I") #raw.index("PART I") #5575
5575
>>> raw=raw[5575:1158049] #粒度是字符
>>> raw.find("PART I") #0
0
```

方法 find() 得到字符串切片需要用到的索引值(raw.index())，用这个切片重新给 raw 赋值，所以现在 raw 以 "PART I" 开始一直到不包含标记内容结尾的句子。

3.1.2 HTML 处理

网络上的文本大部分是 HTML 文件的形式，可以使用网络浏览器将网页作为文本保存为本地文件，然后按照文件方式来访问它。最简单的办法是直接让 Python 来做这份工作，使用 urlopen，输入 html 便可以看到 HTML 的全部内容。

```
C:\Anaconda3>python
Python 3.4.1 |Anaconda 2.1.0 (64-bit)| (default, Sep 24 2014,
18:32:42) [MSC v.1600 64 bit (AMD64)] on win32
Type "help", "copyright", "credits" or "license" for more information.
>>> from urllib.request import urlopen
>>> url="http://poj.org/problem?id=1000"
>>> html=urlopen(url).read()
>>> print(html[:60])
b'<html><head><meta http-equiv="Pragma" content="no-cache"><meta http-equiv="Content-Type" content="text/html; charset=utf-8"><meta http-equiv="Content-Language" content="en-US"><title>1000 -- A+B Problem</title>...
```

从 HTML 中提取文本是常见的任务，将 HTML 字符串作为参数，返回原始文本，然后可以对原始文本进行分词，获得文本结构。

```
>>> from bs4 import BeautifulSoup
>>> from nltk import word_tokenize, Text
>>> raw=BeautifulSoup(html).get_text()
>>> tokens=word_tokenize(raw)
>>> tokens
['1000', '--', 'A+B', 'ProblemOnline', 'JudgeProblem', ...
```

这里仍然含有不需要的内容，包括网站导航及有关报道等。

3.1.3 本地文件读取

读取本地文件需要使用 Python 内置的 open()函数和 read()方法。

```
>>> open('document.txt', 'rU')
<_io.TextIOWrapper name='document.txt' mode='rU' encoding='cp936'>
>>> print(open('document.txt', 'rU').read().encode('cp936').decode())
```

如果解释器无法找到文件，此时需要检查正试图打开的文件是否在正确的目录中。在 Python 中检查当前目录的代码：

```
from os import listdir
listdir('.')
```

在访问一个文本文件时可能遇到的问题是换行的约定。这个约定因操作系统不同而不同，内置的 open()函数的第二个参数用于控制如何打开文件，open('document.txt', 'rU')，'r'意味着以只读方式打开文件（默认），'U'表示"通用"，是指忽略不同的换行约定。

假设已经打开该文件，有几种方法可以阅读此文件。read()方法创建了一个包含整个文件内容的字符串，可以使用一个 for 循环一次读文件中的一行。

```
fp=open('document.txt', 'rU')
for line in fp:
    print(line.strip().encode('cp936').decode())
```

在这里，我们使用 strip()方法删除输入行结尾的换行符。NLTK 中的语料库文件也可以使用这些方法来访问，只需使用 nltk.data.find()来获取语料库项目的文件名，随后就可以使用上述方式打开和阅读它。

```
from nltk.data import find
path=find('corpora/gutenberg/melville-moby_dick.txt')
#FileSystemPathPointer('C:\\nltk_data\\corpora\\gutenberg\\melville-moby_dick.txt')
raw=open(path, 'rU').read()
raw[:70]
```

3.1.4 input 函数

有时想获取用户与程序交互时输入的文本，可以调用 Python 函数 input()，提示用户输入一行数据，保存用户输入到一个变量后，可以像其他字符串那样操纵它。

```
from nltk import word_tokenize
s=input("输入一些文本:") #对于 Python 2.x，请使用 raw_input()
```

3.1.5 自然语言处理流程

自然语言处理流程：打开一个 URL，读取 HTML 格式的内容，去除标记，选择字符的切片，并分词。是否转换为 nltk.Text 对象是可选择的，可以将所有词汇小写并提取词汇表。

在这条流程后面还有很多操作，要正确理解它，这样有助于明确其中提到的每个变量的类型。使用 type(x)可以找出任一 Python 对象 x 的类型。载入一个 URL 或文件的内容，或者去掉 HTML 标记时，正在处理字符串，也就是 Python 的<str>数据类型。将一个字符串分词，会产生一个词的列表，这是 Python 的<list>数据类型，规范化和排序列表产生其他列表。

```
>>> from nltk import word_tokenize, wordpunct_tokenize
>>> raw=open('document.txt').read().encode('cp936').decode()
>>> tokens=word_tokenize(raw)
>>> type(tokens)
<class 'list'>
>>> tokens
['世界', ',', '你好', '!', 'Hello', ',', 'World', '!']
>>> tokens=wordpunct_tokenize(raw)
>>> words=[w.lower() for w in tokens]
>>> type(words)
<class 'list'>
>>> vocab=sorted(set(words))
>>> type(vocab)
<class 'list'>
>>> words
['世界', ',', '你好', '!', 'hello', ',', 'world', '!']
>>> vocab
['!', ',', 'hello', 'world', '世界', '你好']
>>> raw
'世界, 你好!\nHello, World!'
```

3.2 字符串

一个文件的内容在编程语言中是由一个叫作字符串的基本数据类型来表示的。本节将详细探讨字符串，并展示字符串与词汇、文本和文件之间的联系。

3.2.1 字符串的基本操作

可以使用单引号、双引号或三引号来指定字符串，如下面的代码所示。如果一个字符串中包含一个单引号，则必须在单引号前加反斜杠\，让 Python 知道这是字符串中的单引号，也可以将这个字符串放入双引号中（或三引号），否则，字符串内的单引号将被解释为字符串结束标志，Python 解释器会报告语法错误。

```
monty='Monty Python'
circus="Monty Python's Flying Circus"
circus='Monty Python\'s Flying Circus'
circus='''Monty Python's Flying Circus'''
```

有时字符串跨好几行，Python 提供了多种方式来表示它们。在下面的例子中，一个包含两个字符串的序列被连接为一个字符串，我们需要使用反斜杠\或者括号（前面用作转义，这里用作连接），这样解释器就知道第一行的表达式不完整。

```
>>> couplet="Shall I compare thee to a Summer's day?" \
... "Thou are more lovely"
>>> couplet
"Shall I compare thee to a Summer's day?Thou are more lovely"
>>> couplet="Shall I compare thee to a Summer's day?"
>>> "Thou are more lovely"
'Thou are more lovely'
>>> couplet
"Shall I compare thee to a Summer's day?"
>>> couplet=("Shall I compare thee to a Summer's day?"
... "Thou are more lovely")
>>> couplet
"Shall I compare thee to a Summer's day?Thou are more lovely"
```

上述方法没有展现两行之间的换行\n，为此，我们可以使用如下所示的三重引号的字符串。

```
>>> couplet='''Rough winds do shake the darling buds of May,
... And Summer's lease hat hall too short a date:'''
>>> couplet
"Rough winds do shake the darling buds of May, \nAnd Summer's lease hat hall too short a date:"
>>> print(couplet)
Rough winds do shake the darling buds of May,
And Summer's lease hat hall too short a date:
```

现在可以定义字符串，也可以在上面尝试一些简单的操作。首先，让我们来看看+操作，被称为连接。此操作产生一个新字符串，它由两个原始字符串首尾相连粘贴复制而成。可以在词汇之间插入空格，甚至可以对字符串使用乘法操作。

```
'very'+'very'+'very'
'very'*3
>>> 'very'+'very'+'very'
'veryveryvery'
>>> 'very'*3
'veryveryvery'
```

3.2.2 访问单个字符

当索引一个字符串时，可以得到它的一个字符，一个单独的字符并没有什么特别，它只是一个长度为 1 的字符串。与列表一样，如果尝试访问一个超出字符串范围的索引，会得到一个错误。可以使用字符串的负数索引，其中-1 是最后一个字符的索引。当一个字符串的长度为 10 时，索引 5 和 - 5 都指示相同的字符。

```
>>> s='0123456789'
>>> len(s)
10
>>> s[5] #正向，索引从 0 开始
'5'
>>> s[-5] #反向，索引从-1 开始
'5'
```

可以写一个 for 循环，遍历字符串中的字符。

```
sent='colorless green ideas sleep furiously'
for char in sent:
    print(char, end='')
```

也可以计数单个字符，通过将所有字符小写来忽略大小写的区分，并过滤掉非字母字符。

```
C:\Anaconda3>python
Python 3.4.1 |Anaconda 2.1.0 (64-bit)| (default, Sep 24 2014, 18:32:42) [MSC v.1600 64 bit (AMD64)] on win32
Type "help", "copyright", "credits" or "license" for more information.
>>> from nltk import FreqDist
>>> from nltk.corpus import gutenberg
>>> raw=gutenberg.raw('melville-moby_dick.txt')
>>> fdist=FreqDist([ch.lower() for ch in raw if ch.isalpha()])
>>> fdist.keys()
dict_keys(['d', 'z', 'i', 'u', 'n', 'h', 'r', 'b', 'e', 'y', 'c', 'v', 'f', 'm', 'l', 'p', 's', 'q', 'j', 'g', 't', 'w', 'o', 'k', 'a', 'x'])
>>> fdist.items()
dict_items([('d', 38219), ('z', 632), ('i', 65434), ('u', 26697), ('n', 65617), ('h', 62896), ('r', 52134), ('b', 16877), ('e', 117092), ('y', 16872), ('c', 22507), ('v', 8598), ('f', 20833), ('m', 23277), ('l', 42793), ('p', 17255), ('s', 64231), ('q', 1556), ('j', 1082), ('g', 20820), ('t', 87996), ('w', 22222), ('o', 69326), ('k', 8059), ('a', 77916), ('x', 1030)])
>>> fdist.plot()
```

用 fdist.plot()可视化这个分布，一个文本相关的字母频率特征如图 3-1 所示，可以用在文本语言自动识别中。

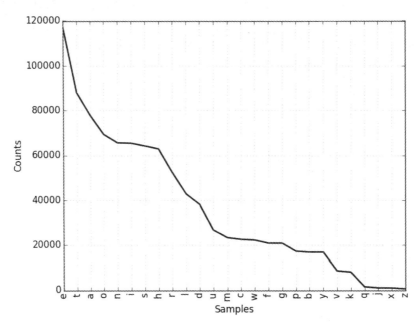

图 3-1　一个文本相关的字母频率特征

3.2.3　子串访问

子串是从一个字符串中取出的任意连续片段，可以很容易地访问子串。切片[m:n]包含从位置 m 到 n-1 中的字符，切片开始于第一个索引，但结束于最后一个索引的前一个；也可以使用负数索引切片，也是同样的规则，从第一个索引开始到最后一个索引的前一个结束，在这里是在空格字符前结束。如果省略了第一个值，子串将从字符串的开头开始；如果省略了第二个值，则子字符串直到字符串的结尾结束。

使用 in 操作符测试一个字符串是否包含（不是属于）一个特定的子串，代码如下：

```
phrase='And now for something completely different'
if 'thing' in phrase:
    'found "thing"'
```

也可以使用 find()找到一个子字符串在字符串内的位置。

```
>>> monty='Monty Python'
>>> monty.find('Python')
6
>>> monty.index('Python')
```

```
6
>>> sent='She sell sea shell by the sea shore .'
>>> sent[4:8]
'sell'
>>> from nltk import word_tokenize
>>> sent='She sell sea shell by the sea shore .'
>>> sent=word_tokenize(sent)
>>> sent[1]
'sell'
```

3.2.4 有用的字符串方法

Python 对处理字符串的支持很全面,s.join(text)连接字符串 s 与 text 中的词汇;s.splitlines()将 s 按行分割成字符串列表。

```
>>> ' '.join(['Hello', ',', 'world', '!'])
'Hello , world !'
>>> 'line1\nline2\n'.splitlines(True)
['line1\n', 'line2\n']
>>> 'line1\nline2\n'.splitlines(False)
['line1', 'line2']
```

3.2.5 列表与字符串的差异

字符串和列表都是一种序列,可以通过索引获取它们中的一部分,可以给它们切片,可以使用连接将它们合并在一起,但是,字符串和列表之间不能连接。

在一个 Python 程序中打开并读入一个文件,此时得到一个对应整个文件内容的字符串。如果使用一个 for 循环来处理这个字符串元素,则所有可以挑选出的只是单个的字符,无法选择粒度(不灵活)。

相比之下,列表中的元素可以很大也可以很小,例如,它们可能是段落、句子、短语、单词、字符。所以,列表的优势是可以灵活地决定它包含的元素,相应的后续处理也变得灵活。因此,在一段 NLP 代码中可能做的第一件事情就是将一个字符串分词放入一个字符串列表中。相反,要将结果写入一个文件或终端时,通常会将它们格式化为一个字符串。列表与字符串的功能有很大不同,列表增加了可以改变其中的元素的能力。

字符串是不可变的,一旦创建了一个字符串,就不能改变它。然而,列表是可变的,其内容可以随时修改。列表支持修改原始值的操作,而不是产生一个新的值。

3.3 使用 Unicode 进行文字处理

3.3.1 Unicode 简介

Unicode 支持超过 100 万种字符，每个字符分配一个编号，称为编码点。在 Python 中，编码点写作\u××××的形式，其中××××是四位十六进制形式数。在一个程序中，可以像普通字符串那样操纵 Unicode 字符串，然而，当 Unicode 字符存储在文件或在终端上显示时，它们必须被编码为字节流。一些编码（如 ASCII），每个编码点使用单字节，所以它们只支持 Unicode 的一个小的子集，其他的编码（如 UTF-8），使用多个字节，可以表示全部的 Unicode 字符。文件中的文本都是有特定编码的，所以需要一些机制来将文本翻译成 Unicode，翻译成 Unicode 的过程叫作解码。相对地，要将 Unicode 写入一个文件或终端时，首先需要将 Unicode 转化为合适的编码，这种将 Unicode 转化为其他编码的过程叫作编码。从 Unicode 的角度来看，字符是可以实现一个或多个字形的抽象实体，只有字形可以出现在屏幕上或被打印在纸上。一个字体是一个字符到字形映射。

3.3.2 从文件中提取已编码文本

读写中文文件，如果不考虑编码，此时不可以正常从文件中读写中文，也无法正常打印中文字段到屏幕上。

```
C:\Anaconda3>python
Python 3.4.1 |Anaconda 2.1.0 (64-bit)| (default, Sep 24 2014,
18:32:42) [MSC v.1600 64 bit (AMD64)] on win32
Type "help", "copyright", "credits" or "license" for more information.
SRC_PATH='src.txt' #src.txt, 编码 UTF-8
DST_PATH='dst.txt'
src_file=open(SRC_PATH, 'r')
dst_file=open(DST_PATH, 'w')
for line in src_file.readlines():
    dst_file.writelines(line)
    print(line.strip())

Traceback (most recent call last):
  File "<stdin>", line 1, in <module>
```

```
UnicodeDecodeError: 'gbk' codec can't decode byte 0xad in position
8: illegal multibyte sequence
>>> src_file.close()
>>> dst_file.close()

SRC_PATH='src.txt'
DST_PATH='dst.txt'
src_file=open(SRC_PATH, 'r', encoding='utf-8')
dst_file=open(DST_PATH, 'w')
for line in src_file.readlines():
    dst_file.writelines(line)
    print(line.strip())

src_file.close()
dst_file.close()
```

查看系统编码的具体转换：

```
import sys
sys.getdefaultencoding()


>>> print(u'中文')
中文
>>> print(u'中文'.encode('gbk'))
b'\xd6\xd0\xce\xc4'
>>> print(u'中文'.encode('gb18030'))
b'\xd6\xd0\xce\xc4'
>>> print('中文')
中文
>>> print(u'中文'.encode('utf-8'))
b'\xe4\xb8\xad\xe6\x96\x87'
```

3.4 使用正则式检测词组搭配

3.4.1 使用基本的元字符

使用正则式ed$查找以ed结尾的词汇，函数re.search(p, s)检查字符串s中是否有模式p，此时需要指定ed，然后使用$，用来匹配单词的末尾。

```
>>> from re import search
>>> search('ed$', 'dejected')
<_sre.SRE_Match object; span=(6, 8), match='ed'>
>>> search('ed$', 'deject')
>>> search('ed$', 'dejected').group()
'ed'
>>> search('ed$', 'dejected').group(0)
'ed'
```

通配符"."匹配任何单个字符。假设有一个由 8 个字母组成的词的字谜室，j 是第 3 个字母，t 是第 6 个字母，空白单元格中的每个地方用一个句点。

```
>>> wordlist='The dejected expression on the face of the loser spoil my victory.This is an abjective idea really.'.split()
>>> [w for w in wordlist if search('^..j..t..$', w)]
['dejected']
```

插入符号"^"匹配字符串的开始，就像"$"符号匹配字符串的结尾，如果不用这两个符号，刚才的例子我们会得到什么样的结果？

```
>>> [w for w in wordlist if search('..j..t..', w)]
['dejected', 'abjective']
```

符号"?"表示前面的字符是可选的，因此^e-?mail$将匹配 email 和 e-mail。可以使用 sum(1 for w in text if re.search('^e-?mail$', w))计数一个文本中这个词出现的总次数。

```
>>> text=['email', 'e-mail', 'e', 'mail']
>>> sum(1 for w in text if search('^e-?mail$', w))
2
>>> [1 for w in text if search('^e-?mail$', w)]
[1, 1]
```

3.4.2 范围与闭包

T9 系统如图 3-2 所示，它主要用于在手机上输入文本。

图 3-2 T9 系统

两个或两个以上的词汇以相同的击键顺序输入，这叫作输入法联想提示。例如，hole 和

golf 都是通过输入序列 4653。还有哪些其他词汇由相同的序列产生呢？这里使用正则式 ^[ghi][mno][jlk][def]$：

```
>>> from nltk.corpus import words
>>> from re import search
>>> wordlist=words.words('en')
>>> [w for w in wordlist if search('^[ghi][mno][jlk][def]$', w)]
['gold', 'golf', 'hold', 'hole']
```

表达式的第一部分^[ghi]匹配以 g、h 或 i 开始的词，表达式的第二部分[mno]限制了第二个字符是 m、n 或 o，第三部分和第四部分也是限制，四个字母要满足所有这些限制。注意，方括号内字符的顺序是没有关系的（集合无序性），所以可以写^[ghi][mno][jkl][def]$匹配同样的词汇。如果不用正则式呢？用 match，findall 也可以。

让我们进一步探索"+"符号，它可以适用于单个字母或括号内的字母集。"+"和"*"符号有时被称为 Kleene 闭包。

运算符"^"出现在方括号内的第一个字符位置时有其他功能。例如，[^aeiouAEIOU]匹配除元音字母之外的所有字母。

下面是另外一些正则式的例子，用来寻找匹配特定模式的词汇标识符，这些例子演示如何使用一些新的符号：\、{}、()和|。

```
>>> from nltk.corpus import treebank
>>> from re import search
>>> wsj=sorted(set(treebank.words()))
>>> [w for w in wsj if search('^[0-9]+\.[0-9]+$', w)]
['0.0085', '0.05', ..., '96.4', '98.3', '99.1', '99.3']
>>> [w for w in wsj if search('^[0-9]{4}$', w)]
['1614', '1637', '1787', '1901', '1903', '1917', '8300']
>>> [w for w in wsj if search('(ed|ing)$', w)]
['62%-owned', 'Absorbed', 'According', 'Adopting', ...]
```

反斜杠表示其后面的字母不再有特殊的含义，而是按照字面的表示匹配词中特定的字符，因此，虽然"."有特殊含义，但"\."只匹配一个句号。大括号表达式，如{3, 5}，表示前面的项目重复指定次数。管道字符|表示从其左边的内容和右边的内容中选择一个。圆括号表示一个操作符()的范围，它们可以与管道（或叫析取）符号一起使用，如 w(i|e|ai|oo)t，匹配 wit、wet、wait 和 woot。

可以省略这个例子里的最后一个表达式中的括号，使用 ed|ing$搜索看看会发生什么。

```
len([w for w in wsj if re.search('(ed|ing)$', w)]) #1841, 相当于 ed$, ing$
len([w for w in wsj if re.search('ed|ing$', w)]) #1969, 相当于 ed, ing$
len([w for w in wsj if re.search('[ed|ing]$', w)]) #4618 相当于[e, d, |, i, n, g]
```

对 Python 解释器而言，一个正则式与任何其他字符串没有区别，如果字符串中包含一个反斜杠后面跟一些特殊字符，Python 解释器将会特殊处理它们。例如，\b 会被解释为一个退格符号。一般情况下，当使用含有反斜杠的正则式时，我们应该告诉解释器一定不要解释字符串里面的符号，而仅仅是将它直接传递给 re 库来处理，我们通过给字符串加一个前缀 r 来表明它是一个原始字符串。例如，原始字符串 r'\band\b'包含两个\b 符号会被 re 库解释为匹配词的边界而不是解释为退格字符。

```
>>> r'\band\b'
'\\band\\b' #r 加了一个\
>>> print(r'\band\b') #Python 解释\\为一个\
\band\b
>>> '\band\b' #Python 解释\b 为一个\x08，即退格字符
'\x08and\x08'
>>> from re import search
>>> wordlist=['and', 'youandme']
>>> [w for w in wordlist if search(r'\band\b', w)]
```

r 处理得到'\\band\\b'。在 Python 中，第一个\是转义符，将第二个\转为普通符号，得到'\band\b'，Python 传递'\band\b'给 re 库处理。在 re 库中，\b 匹配词的边界，即 w 完全匹配 and。

```
['and']
>>> [w for w in wordlist if search('\band\b', w)]
```

在 Python 中，第一个\是转义符，将第二个 b 转为\x08，即退格字符，得到'\x08and\x08'，Python 传递'\x08and\x08'给 re 库处理，匹配首尾为退格字符的 and，显然没有。

```
[]
```

3.5　正则式的应用

使用 re.search(regexp，w)匹配一些正则式 regexp 来搜索词 w，除了检查一个正则式是否匹配一个单词外，还可以使用正则式从词汇中提取特征或以特殊的方式来修改词。

3.5.1　提取字符块

通过 findall()方法找出所有匹配指定正则式的字符块，让我们找出一个词中的元音，再进行计数。

```
>>> from re import findall, search
>>> word='supercalifragilisticexpialidocious' #欢乐满人间；超级长字；难以置信的
>>> findall('[aeiou]', word)
```

```
['u', 'e', 'a', 'i', 'a', 'i', 'i', 'i', 'e', 'i', 'a', 'i', 'o',
'i', 'o', 'u']
>>> len(findall('[aeiou]', word))
16
>>> word='supercalifragilisticexpialidocious'
>>> findall('(a|e|i|o|u)', word)
['u', 'e', 'a', 'i', 'a', 'i', 'i', 'i', 'e', 'i', 'a', 'i', 'o',
'i', 'o', 'u']
```

不用正则表达式：

```
>>> word='supercalifragilisticexpialidocious'
>>> [c for c in word if c in 'aeiou']
['u', 'e', 'a', 'i', 'a', 'i', 'i', 'i', 'e', 'i', 'a', 'i', 'o',
'i', 'o', 'u']
>>> word='supercalifragilisticexpialidocious'
>>> search('(a|e|i|o|u)', word).group()
'u'
>>> search('(a|e|i|o|u)', word).groups()
('u', )
>>> from nltk import re_show
>>> word='supercalifragilisticexpialidocious'
>>> re_show('(a|e|i|o|u)', word)
s{u}p{e}rc{a}l{i}fr{a}g{i}l{i}st{i}c{e}xp{i}{a}l{i}d{o}c{i}{o}{u}s
```

让我们来看看一些文本中的两个或两个以上的元音序列，并确定它们的相对频率。

```
>>> from nltk import FreqDist
>>> from re import findall
>>> wsj=['The', 'buildings', 'are', 'really', 'magnificent', '.']
>>> fd=FreqDist([vs for word in wsj for vs in findall('[aeiou]{2, }', word)])
>>> fd.items()
dict_items([('ea', 1), ('ui', 1)])

>>> word='modification'
>>> findall('[aeiou]{2, }', word)
['io']
>>> wsj=['The', 'buildings', 'are', 'really', 'magnificent', '.']
```

```
>>> fd=FreqDist(vs for word in wsj for vs in findall('[aeiou]{2,}',
word))
>>> fd.items()
dict_items([('ea', 1), ('ui', 1)])
>>> wsj=['The', 'buildings', 'are', 'really', 'magnificent', '.']
>>> x=[]
>>> for word in wsj:
...     y=findall('[aeiou]{2,}', word) #注意,{2,}中不能有空格符号。
...     for vs in y:
...         x.append(vs)
...
>>> fd=FreqDist(x)
>>> fd.items()
dict_items([('ea', 1), ('ui', 1)])
>>> wsj=['The', 'buildings', 'are', 'really', 'magnificent', '.']
>>> x=[]
>>> for word in wsj:
...     y=findall(r'[aeiou]{2,}', word)
...     x.extend(y)
...
>>> fd=FreqDist(x)
>>> fd.items()
dict_items([('ea', 1), ('ui', 1)])
```

用正则式替换下面 Python 代码中的?,将字符串'2009-12-31'转换为一个整数列表[2009,12,31]。

```
C:\Anaconda3>python
Python 3.4.1 |Anaconda 2.1.0 (64-bit)| (default, Sep 24 2014,
18:32:42) [MSC v.1600 64 bit (AMD64)] on win32
[int(n) for n in findall(?__, '2009-12-31')]
>>> from re import findall, search
>>> [int(n) for n in findall('\d+', '2009-12-31')]
#注意,Python 3.4.1可以省去前缀r!
[2009, 12, 31]
>>> [int(n) for n in search('(\d+)-(\d+)-(\d+)', '2009-12-31').groups()]
[2009, 12, 31]
>>> [int(n) for n in '2009-12-31'.split('-')]
[2009, 12, 31]
```

3.5.2 单词处理

英文文本是高度冗余的，忽略掉词内部的元音仍然可以很容易地阅读，有些时候这很明显，例如，declaration 变成 dclrtn，inalienable 变成 inlnble，保留所有词首或词尾的元音序列，正则式匹配词首元音序列、词尾元音序列和所有的辅音，其他被忽略。这三个阶段从左到右处理，如果词匹配了三个部分之一，正则式后面的部分将被忽略。使用 findall() 提取所有匹配词中的字符，随后使用".join()将它们连接在一起。

```
>>> from re import findall
>>> from textwrap import fill
>>> regexp='^[AEIOUaeiou]+|[AEIOUaeiou]+$|[^AEIOUaeiou]'
>>> def compress(word):
...     pieces=findall(regexp, word)
...     return ''.join(pieces)
#用空串连接在一起形成词
...
>>> txt='Whenever you see a variable, look for a my declaration higher up in the same block. '.split()
>>> print(fill(' '.join(compress(w) for w in txt)))
#词用空格串连接在一起形成句
Whnvr you see a vrbl, lk fr a my dclrtn hghr up in the sme blck.

from re import findall
regexp='^[AEIOUaeiou]+|[AEIOUaeiou]+$|[^AEIOUaeiou]'
#|左右两边不能有空格符
>>> word='buildings'
>>> pieces=findall(regexp, word)
>>> pieces
['b', 'l', 'd', 'n', 'g', 's']
>>> word='declaration'
>>> pieces=findall(regexp, word)
>>> pieces
['d', 'c', 'l', 'r', 't', 'n']
>>> word='inalienable'
>>> pieces=findall(regexp, word)
>>> pieces
['i', 'n', 'l', 'n', 'b', 'l', 'e']
```

findall 匹配 buildings 词首元音序列（没有）、词尾元音序列（没有）和词中所有的辅音，pieces=['b', 'l', 'd', 'n', 'g', 's']，使用".join(pieces)将它们连接在一起'bldngs'。

从 rotas.dic 中提取所有辅音元音序列，如 ka 和 si。因为每部分都是成对的，所以它可以被用来初始化一个条件频率分布，然后为每对的频率制表。

```
>>> from re import findall  #导入findall
>>> from nltk.corpus import toolbox  #解压toolbox.zip
>>> from nltk import ConditionalFreqDist
>>> ws = toolbox.words('rotokas.dic')  #加上ok, toolbox.entries
>>> cvs = [cv for w in ws for cv in findall('[ptksvr][aeiou]', w)]
>>> cvs
['ka', 'ka', 'ka', 'ka', 'ka', 'ro', 'ka', 'ka', 'vi', ...]
>>> cfd = ConditionalFreqDist(cvs)
>>> cfd.tabulate()
    a    e    i    o    u
k 418  148   94  420  173
p  83   31  105   34   51
r 187   63   84   89   79
s   0    0  100    2    1
t  47    8    0  148   37
v  93   27  105   48   49
```

查看 s 行和 t 行，它们是部分的互补分布。这个证据表明它们不是这种语言中的不同的音素，从而我们可以从罗托卡特语字母表中去除 s，简单加入一个发音规则，当字母 t 后跟 i 时发 s 的音。如果想要检查表格中数字背后的词汇，有一个索引允许我们迅速找到包含一个给定的辅音元音对的单词的列表将会有帮助。

```
>>> from nltk import Index
>>> cv_word_pairs = [(cv, w) for w in ws for cv in findall('[ptksvr][aeiou]', w)]
>>> [(cv, w) for w in ws for cv in findall('[ptksvr][aeiou]', w)]
[('ka', 'kaa'), ('ka', 'kaa'), ('ka', 'kaa'), ('ka', 'kaakaaro'), ...]
>>> cv_index = Index(cv_word_pairs)
>>> cv_index
defaultdict(<class 'list'>, {'vo': ['kaakaavo', 'kakavoro', ...]})
>>> cv_index['su']  #['kasuari']  #所有含有su的词汇。
['kasuari']
cv_index['ka']
cv_index['ri']
```

这段代码依次处理每个词 w，对每一个词找出匹配正则式[ptksvr][aeiou]的所有子字符串。对于词 kasuari，它找到 ka、su 和 ri，从而列表 cv_word_pairs 将包含('ka', 'kasuari')、('su', 'kasuari')和('ri', 'kasuari')。

3.5.3 查找词干

在使用网络搜索引擎时，通常不介意文档中词汇与搜索条件的后缀形式是否相同，查询 laptops 会找到含有 laptop 的文档，反之亦然。事实上，laptop 与 laptops 只是字典中的同一个词或词条的单复数形式。

对于一些语言处理任务，我们想忽略词语结尾，只处理词干。抽出词干的方法有很多种，直接去掉任何看起来像一个后缀的字符，是一种简单、直观的方法。

```
>>> def stem(word):
...     global j
...     for suffix in ['ing', 'ly', 'ed', 'ious', 'ies', 'ive', 'es', 's', 'ment']:
...         j=j+1
...         if word.endswith(suffix):
...             return word[:-len(suffix)]
...     return word
...
>>> word='laptops'
>>> j=0
>>> stem(word)
'laptop'
>>> print(j)
8
>>> word='processing'
>>> j=0
>>> stem(word)
'process'
>>> print(j)
1
```

3.5.4 搜索已分词文本

可以使用一种特殊的正则式搜索一个文本中多个词，这里的文本是一个词列表 Text，例如<a><man>，找出文本中所有 a man 的实例，尖括号用于标记词的边界\b，尖括号之间的所有空白都被忽略。它只对 NLTK 中的 findall()方法处理文本有效，对 re 中的 findall()方法处理文本无效。在下面的<a>(<.*>)<man>例子中，使用<.*>将匹配所有单个标识符，将它括在括号里，于是只匹配词，例如 monied，而不匹配短语，例如 a monied man。

```
>>> from nltk import Text
>>> from nltk.corpus import gutenberg, nps_chat
>>> moby=Text(gutenberg.words('melville-moby_dick.txt'))
>>> moby.findall("<a>(<.*>)<man>") #similar
monied; nervous; dangerous; white; white; white; pious; good;
mature; white; Cape; great; wise; wise; butterless; ...
```

good 是形容词，butterless 是形容词，上下文相同。

下面巩固对正则式模式与替换的理解。使用 re_show(p，s)能标注字符串 s 中所有匹配模式 p 的地方。

```
>>> from nltk import re_show
>>> p='[A-Z][a-z]*'
>>> re_show(p, 'Price')
{Price}
>>> p='\d+(\.\d+)?'
>>> re_show(p, '2312.12345')
{2312.12345}
```

当我们研究的语言现象与特定词语相关时，建立搜索模式是很容易的。在某些情况下，一个小小的创意可能会花很大工夫，例如，在大型文本语料库中搜索 x and other ys 形式的表达式能发现上位词。

```
>>> from nltk.corpus import brown
>>> from nltk import Text
>>> hobbies_learned=Text(brown.words(categories=['hobbies','learned']))
>>> hobbies_learned.findall("<\w*><and><other><\w*s>") #re.findall()/
speed and other activities; water and other liquids; tomb and other
landmarks; Statues and other monuments; pearls and other jewels; charts
and other items; roads and other features; figures and other objects;
military and other areas; demands and other factors;...
```

结果 water and other liquids 表明 water（水）是类型 liquids（液体）的一个实例（上位词）。只要有足够多的文本，这种做法会给我们一整套有用的分类标准信息，而不需要任何手动操作。然而，搜索结果中通常会包含误报，即我们想要排除的情况。例如，结果 demands and other factors 表明 demands 是类型 factor 的一个实例（上位词），但是这句话实际上是关于要求增加工资的。尽管如此，我们仍可以通过手工纠正这些搜索的结果来构建自己的英语概念本体。

搜索语料也会有遗漏的问题,即漏掉了我们想要包含的情况。仅仅因为我们找不到任何一个搜索模式的实例,就断定一些语言现象在一个语料库中不存在,是有问题的,也许我们只是没有足够仔细地思考合适的模式。查找模式 as x as y 的实例以发现实体及其属性信息。

```
>>> from nltk.corpus import brown
>>> hobbies_learned=Text(brown.words(categories=['hobbies','learned']))
>>> hobbies_learned.findall("<as><\w*><as><\w*>")
as accurately as possible; ...; as soon as possible; ...
```

3.6 规范化文本

3.6.1 词干提取器

NLTK 中包含一些现成的词干提取器,如果需要一个词干提取器,应该优先使用它们中的一个,而不是使用正则式制作自己的词干提取器。因为 NLTK 中的词干提取器能处理的不规则的情况很多,例如,PorterStemmer 和 LancasterStemmer 词干提取器能按照它们自己的规则剥离词缀。PorterStemmer 词干提取器正确处理了词 lying,将它映射为 lie(正则式却将它映射为 ly),而 LancasterStemme 词干提取器并没有处理好。

```
>>> from nltk import word_tokenize, PorterStemmer, LancasterStemmer
>>> raw="""DENNIS: Listen, strange women lying in ponds distributing
swords is no basis for a system of government. Supreme executive power
derives from a mandate from the masses, not from some farcical aquatic
ceremony."""
>>> tokens=word_tokenize(raw)
>>> porter=PorterStemmer()
>>> lancaster=LancasterStemmer()
>>> [porter.stem(t) for t in tokens]
['DENNI', ':', 'Listen', ',', 'strang', 'women', 'lie', 'in', 'pond',
'distribut', 'sword', 'is', 'no', 'basi', 'for', 'a', 'system', 'of',
'govern', '.', 'Suprem', 'execut', 'power', 'deriv', 'from', 'a', 'mandat',
'from', 'the', 'mass', ',', 'not', 'from', 'some', 'farcic', 'aquat',
'ceremoni', '.']
>>> [lancaster.stem(t) for t in tokens]
['den', ':', 'list', ',', 'strange', 'wom', 'lying', 'in', 'pond',
'distribut', 'sword', 'is', 'no', 'bas', 'for', 'a', 'system', 'of', 'govern',
'.', 'suprem', 'execut', 'pow', 'der', 'from', 'a', 'mand', 'from', 'the',
'mass', ',', 'not', 'from', 'som', 'farc', 'aqu', 'ceremony', '.']
```

词干提取过程没有明确定义，我们通常选择最适合我们应用的词干提取器。如果要索引一些文本和使搜索支持不同词汇形式，PorterStemmer 词干提取器是一个很好的选择。注意：tokens.index('is')、raw.index('is')与 Index()不同。

使用词干提取器索引文本。

```
from nltk import Index
class IndexedText(object):
    def __init__(self, stemmer, text):
        self._text=text
        self._stemmer=stemmer
        self._index=Index((self._stem(word), i) for i, word in enumerate(text))
    def concordance(self, word, width=40): #不是有内置的吗？
        key=self._stem(word)
        wc=width/4 #words of context
        for i in self._index[key]:
            lcontext=' '.join(self._text[i-wc:i])
            rcontext=' '.join(self._text[i:i+wc])
            ldisplay='%*s' % (width, lcontext[-width:]) #*, 输出宽度参数
            rdisplay='%-*s' % (width, rcontext[:width]) #-, 左对齐
            print(ldisplay, rdisplay)
    def _stem(self, word):
        return self._stemmer.stem(word).lower()
from nltk import PorterStemmer
porter=PorterStemmer()
from nltk.corpus import webtext
grail=webtext.words('grail.txt') #grail
text=IndexedText(porter, grail)
text.concordance('lie')
```

测试：

```
import nltk
class IndexedText(object):
    def __init__(self, stemmer, text):
        self._text=text
        self._stemmer=stemmer
        self._index=nltk.Index((self._stem(word), i) for i, word in enumerate(text))
    def concordance(self, word, width=8):  #不是有内置的吗？
```

```
            key=self._stem(word)
            wc=width/4 # words of context
            for i in self._index[key]:
                lcontext=' '.join(self._text[i-wc:i])
#取中心词左边，即上文 wc=width/4=2 个词以空格分开构成一个串，数据粒度变为字符
                rcontext=' '.join(self._text[i:i+wc])
                ldisplay='%*s' % (width, lcontext[-width:])
#从右往左，取中心词左边 width=8 个字符，即上文右对齐，即往中心词靠拢，宽度为
width=8 字符，不足的左边补空格显示。-width=-8 表示从右往左数 8 的位置作为起始位置
                rdisplay='%-*s' % (width, rcontext[:width])
                print ldisplay, rdisplay
    def _stem(self, word):
        return self._stemmer.stem(word).lower()

porter=nltk.PorterStemmer()
grail=nltk.corpus.webtext.words('test.txt')
#text=nltk.Text(grail) #???????
#text.concordance('takes')  # ? 不支持不同词汇形式
'''搜索单词 word 在 text 中出现的次数，并显示每次出现时的上下文。
注意，搜索的词在每一行都是对齐的，不区分大小写，是全字匹配'''
# C:\nltk_data\corpora\webtext\test.txt
text=IndexedText(porter, grail)
text.concordance('takes')
```

按单词索引文本，支持不同词汇形式：

```
She was too tired to take a shower.
This takes me back.
↓words 输入文本并分词
['She', 'was', 'too', 'tired', 'to', 'take', 'a', ...]
↓enumerate 显化并记住单词原位置，后面即将按单词索引(字典)而不是原位置索引
(列表)，但两者是有联系的。
 [(0, 'She'), (1, 'was'), (2, 'too'), (3, 'tired'), (4, 'to'), (5,
'take'), (6, 'a'), (7, 'shower'), (8, '.'), (9, 'This'), (10, 'takes'),
(11, 'me'), (12, 'back'), (13, '.')]
 ↓lower, stem, 规范化，如 Take 和 take、takes 和 take 在字典里是一个词。词
干提取并将单词作为关键词，如(5, 'take')→('take', 5), (10, 'takes') →
('take', 10)
 [('she', 0), ('wa', 1), ('too', 2), ('tire', 3), ('to', 4), ('take',
5), ('a', 6), ('shower', 7), ('.', 8), ('thi', 9), ('take', 10), ('me',
11), ('back', 12), ('.', 13)]
```

↓Index 按单词索引，如('take',5)、('take',10)按 take 索引为'take':[5, 10]，5 和 10 是类 take 在原文中位置索引，这里全部收集起来构成一个单链表(通过列表实现)。现在就可以通过输入关键词(如 take)，就可以找到它在原文中的位置 5 和 10，利用这个位置就可以找到其上下文。
{'a': [6], 'me': [11], 'too': [2], 'wa': [1], 'tire': [3], 'back': [12], '.': [8, 13], 'shower': [7], 'to': [4], 'take': [5, 10], 'thi': [9], 'she': [0]})

索引文本：

```
takes
↓lower, stem, 规范化
take
↓_index['take']，按单词搜索，找到在原文的位置
[5, 10]
↓以此为中心，取左右两边 wc=width/4=8/4=2 个单词，假设单词平均长度为 4 个字符。'['tired', 'to'] ['take' 'a']
↓
'tired to' 'take a'
```

测试：

```
>>> from nltk import Text, PorterStemmer
>>> from nltk.corpus import webtext
>>> def normalized(word):
...     porter=PorterStemmer()
...     return porter.stem(word.lower())
...
>>> grail=webtext.words('test.txt')
>>> grail=[normalized(word) for word in grail]
>>> text=Text(grail)
>>> text.concordance(normalized('takes'))   #但查询后文本变了!
Displaying 2 of 2 matches:
           she wa too tire to take a shower . thi take me back .
           take a shower . thi take me back .
```

3.6.2 词形归并

WordNet 词形归并器删除词缀产生的词都是在它的字典中的词。要做到这一点，就必须要有额外的检查过程，这个额外的检查过程使词形归并器比刚才提到的词干提取器要慢。言下之意，词干提取器产生的词有可能不在它的字典中，它并没有处理 lying，但它将 women 转换为 woman。

```
>>> from nltk import word_tokenize, WordNetLemmatizer
>>> raw="""DENNIS: Listen, strange women lying in ponds distributing swords is no basis for a system of government. Supreme executive power derives from a mandate from the masses, not from some farcical aquatic ceremony. """
>>> tokens=word_tokenize(raw)
>>> wnl=WordNetLemmatizer()
>>> [wnl.lemmatize(t) for t in tokens] #['lemətaiz]
['DENNIS', ':', 'Listen', ',', 'strange', 'woman', 'lying', ..., '.']
>>> wnl.lemmatize('lying', pos='v') #lie
'lie'
import nltk
raw="""take takes taking to taken"""
tens=nltk.word_tenize(raw)
wnl=nltk.WordNetLemmatizer()
[wnl.lemmatize(t) for t in tens] #['lemətaiz]
['take', 'take', 'taking', 'to', 'taken']
```

Lemmatization 把一个任何形式的语言词汇还原为一般形式，在标记词性的前提下效果比较好。

```
wnl.lemmatize('lying', pos='v') #lie
```

NLTK 中这个词形还原工具的问题是需要手动指定词性，例如上面例子中的'lying'这个词，如果不加后面 pos 参数，输出的结果将会是'lying'本身。

在实际应用中使用 NLTK 进行词形还原的解决方案：

（1）输入一个完整的句子；
（2）用 NLTK 提供的工具对句子进行分词和词性标注；
（3）将得到的词性标注结果转换为 WordNet 的格式；
（4）使用 WordNetLemmatizer 对词进行词形还原。

其中，分词和词性标注又有数据依赖：

```
from nltk.corpus import wordnet
from nltk import word_tenize, pos_tag
from nltk.stem import WordNetLemmatizer
def get_wordnet_pos(treebank_tag):
    if treebank_tag.startswith('J'):
        return wordnet.ADJ
    elif treebank_tag.startswith('V'):
        return wordnet.VERB
    elif treebank_tag.startswith('N'):
```

```
        return wordnet.NOUN
    elif treebank_tag.startswith('R'):
        return wordnet.ADV
    else:
        return None

def lemmatize_sentence(sentence):
    res = []
    lemmatizer = WordNetLemmatizer()
    for word, pos in pos_tag(word_tenize(sentence)):
        wordnet_pos = get_wordnet_pos(pos) or wordnet.NOUN
        res.append(lemmatizer.lemmatize(word, pos=wordnet_pos))
    return res

sentence='She was too tired to take a shower .'
lemmatize_sentence(sentence)
```

lemmatize 比 stemming 版本更高级，会考虑上下文。

```
e.g. "are, is, being" -> "be" etc.
[('She', 'PRP'), ('was', 'VBD'), ('too', 'RB'), ('tired', 'VBN'),
('to', 'TO'), ('take', 'VB'), ('a', 'DT'), ('shower', 'NN'), ('.', '.')]
['She', 'be', 'too', 'tire', 'to', 'take', 'a', 'shower', '.']
```

如果想编译一些文本的词汇，或者想要一个有效词条（或中心词）列表，可以选择 WordNet 词形归并器。

另一个规范化任务涉及识别非标准词，包括数字、缩写、日期以及任何此类标识符到一个特殊的词汇的映射，例如，每一个十进制数可以被映射到一个单独的标识符 0.0，首字母缩写可以映射为 AAA，这样可以使词汇量变小，提高许多语言建模任务的准确性。

3.7 用正则式为文本分词

3.7.1 分词的简单方法

英文文本分词，一种非常简单的方法是在空格符处分割文本，可以使用 re.split()在空格符处分割原始文本。使用正则式能实现同样的功能，匹配字符串中的所有空格符是不够的，因为这将导致分词结果包含\n 换行符，我们需要匹配任何数量的空格符、制表符或换行符。

```
>>> raw="""'When I'M a Duchess, ' she said to herself,
... (not in a very hopeful tone though),
```

```
... 'I won't have any pepper in my kitchen AT ALL."""
>>> raw.split()
["'When", "I'M", 'a', 'Duchess,', "'", 'she', 'said', 'to', 'herself,
', '(not', 'in', 'a', 'very', 'hopeful', 'tone', 'though),', "'I",
"won't", 'have', 'any', 'pepper', 'in', 'my', 'kitchen', 'AT', 'ALL.']
>>> from re import split
>>> split(' ', raw)
["'When", "I'M", 'a', 'Duchess,', "'", 'she', 'said', 'to', 'herself,
', '\n(not', 'in', 'a', 'very', 'hopeful', 'tone', 'though),', "\n'I",
"won't", 'have', 'any', 'pepper', 'in', 'my', 'kitchen', 'AT', 'ALL.']
>>> split('[ \t\n]+', raw)
["'When", "I'M", 'a', 'Duchess,', "'", 'she', 'said', 'to', 'herself,
', '(not', 'in', 'a', 'very', 'hopeful', 'tone', 'though),', "'I",
"won't", 'have', 'any', 'pepper', 'in', 'my', 'kitchen', 'AT', 'ALL.']
>>> split('\s+', raw)
["'When", "I'M", 'a', 'Duchess,', "'", 'she', 'said', 'to', 'herself,
', '(not', 'in', 'a', 'very', 'hopeful', 'tone', 'though),', "'I",
"won't", 'have', 'any', 'pepper', 'in', 'my', 'kitchen', 'AT', 'ALL.']
```

正则式[\t\n]+匹配一个或多个空格、制表符\t或换行符\n。re 库内置的缩写\s，表示匹配所有空白字符。在空格符处分割文本，如 not 和 herself 这样的标识符；还有一种方法是使用 Python 提供的字符类\w，词中的字符相当于[a-zA-Z0-9_]，定义了这个类的补\W，即所有字母、数字和下划线以外的字符。我们可以在一个简单的正则式中使用\W 来分割所有单词字符以外的输入。

```
>>> split('\W+', raw)
['', 'When', 'I', 'M', 'a', 'Duchess', 'she', 'said', 'to', 'herself',
'not', 'in', 'a', 'very', 'hopeful', 'tone', 'though', 'I', 'won', 't',
'have', 'any', 'pepper', 'in', 'my', 'kitchen', 'AT', 'ALL', '']
```

可以看到，在开始和结尾都给了一个空字符串，通过 findall('\w+', raw)使用模式匹配词汇而不是空白符号，得到相同的标识符，但没有空字符串。正则式\w+|\S\w*先尝试匹配词中字符的所有序列。如果没有找到匹配的，它会尝试匹配后面跟着词中字符的任何非空白字符，\S 是\s 的补。这意味着标点会与跟在后面的字母，如's，在一起，但两个或两个以上的标点字符序列会被分割。

```
>>> from re import findall
>>> findall('\w+', raw)
['When', 'I', 'M', 'a', 'Duchess', 'she', 'said', 'to', 'herself',
'not', 'in', 'a', 'very', 'hopeful', 'tone', 'though', 'I', 'won', 't',
'have', 'any', 'pepper', 'in', 'my', 'kitchen', 'AT', 'ALL']
```

```
>>> findall('\w+|\S\w*', raw)
["'When", 'I', "'M", 'a', 'Duchess', ',', ',', "'", 'she', 'said', 'to',
'herself', ',', '(not', 'in', 'a', 'very', 'hopeful', 'tone', 'though',
')', ',', "'I", 'won', "'t", 'have', 'any', 'pepper', 'in', 'my',
'kitchen', 'AT', 'ALL', '.']
```

扩展前面表达式中的\w+，允许连字符和撇号:\w+([-'])\w+)*。这个表达式表示\w+后面跟零个或更多[-']\w+的实例，它会匹配 hot-tempered 和 it's。我们需要在这个表达式中包含?:，原因在前面已经讨论过。还需要添加一个模式来匹配引号字符让它们与它们包括的文字分开。

```
>>> findall("\w+(?:[-']\w+)* | ' | [-.(]+ | \S\w*", raw)
['When ', "I'M ", 'a ', "'", 'she ', 'said ', 'to ', 'not ', 'in
', 'a ', 'very ', 'hopeful ', 'tone ', 'I ', "won't ", 'have ', 'any
', 'pepper ', 'in ', 'my ', 'kitchen ', 'AT ']
```

在这个例子中的表达式也包括[-.(]+，这会使双连字符--和省略号...被单独分词。

```
>>> findall("[-.(]+ | ' ", " -- ... ( ' ")
['-- ', '... ', '( ']
```

3.7.2 NLTK 的正则式分词器

函数 nltk.regexp_tokenize()与 re.findall()类似，nltk.regexp_tokenize()分词效率更高。为了增强可读性，我们将正则式分为几行写，每行添加一个注释。注意：(?x)标志告诉 Python 去掉嵌入的空白字符和注释。

```
>>> from nltk import regexp_tokenize
>>> text='That U.S.A. poster-print costs $12.40 ...'
>>> pattern='''(?x) #set flag to allow verbose regexps
... (?:[A-Z]\.)+
... |\w+(?:-\w+)*
... |\$?\d+(?:\.\d+)?%?
... |\.\.\.
... |[][.,;"'?():-_`]
... '''
>>> regexp_tokenize(text, pattern)
['That', ' ', 'U.S.A.', ' ', 'poster-print', ' ', 'costs', ' ',
'$12.40', ' ', '...']
text='That U.S.A. poster-print costs $12.40…'
pattern=r'''([A-Z]\.)+|\w+(-\w+)*|\$?\d+(\.\d+)?%?|\.\.\.|[][.,
;"'?():-_`]'''
```

```
regexp_tokenize(text, pattern)
['That', 'U.S.A.', 'poster-print', 'costs', '$12.40']
import re
text='That U.S.A. poster-print costs $12.40...'
pattern=r'''(?:[A-Z]\.)+|\w+(?:-\w+)*|\$?\d+(?:\.\d+)?%?|\.\.\.|[][.,;"'?():-_`]'''
re.findall(pattern, text)
['That', 'U.S.A.', 'poster-print', 'costs', '$12.40']
import re
text='That U.S.A. poster-print costs $12.40…'
pattern=r'''([A-Z]\.)+|\w+(-\w+)*|\$?\d+(\.\d+)?%?|\.\.\.|[][.,;"'?():-_`]'''
re.findall(pattern, text)
[('', '', ''), ('A.', '', ''), ('', '-print', ''), ('', '', ''), ('', '', '.40')]
import nltk
text="""'When I'M a Duchess, ' she said to herself, (not in a very hopeful tone though), 'I won't have any pepper in my kitchen AT ALL. Soup does very well without--Maybe it's always pepper that makes people hot-tempered, '..."""
pattern=r'''(?x) #set flag to allow verbose regexps
    ([A-Z]\.)+ #abbreviations, e.g. U.S.A.
    |\w+(-\w+)* #words with optional internal hyphens
    |\$?\d+(\.\d+)?%? #currency and percentages, e.g. $12.40, 82%
    |\.\.\. #ellipsis
    |[][.,;"'?():-_`] #these are separate tens
    '''
nltk.regexp_tokenize(text, pattern)
["'", 'When', 'I', "'", 'M', 'a', 'Duchess', ',', "'", 'she', 'said', 'to', 'herself', ',', '(', 'not', 'in', 'a', 'very', 'hopeful', 'tone', 'though', ')', ',', "'", 'I', 'won', "'", 't', 'have', 'any', 'pepper', 'in', 'my', 'kitchen', 'AT', 'ALL', '.', 'Soup', 'does', 'very', 'well', 'without', 'Maybe', 'it', "'", 's', 'always', 'pepper', 'that', 'makes', 'people', 'hot-tempered', ',', "'", '...']
```

使用 verbose 标志时，可以不再使用' '来匹配一个空格字符，使用"\s"代替。regexp_tokenize()函数有一个可选的 gaps 参数。设置为 True 时，正则式指定标识符间的距离，就像使用 re.split()一样。

可以使用 set(tokens).difference(wordlist)通过比较分词结果与一个词表，然后报告没有在词表出现的标识符，来评估一个分词器。

3.8 分割

3.8.1 断句

在词级水平处理文本时通常假定能够将文本划分成单个句子。正如我们已经看到，一些语料库已经提供在句子级别的访问，在下面的例子中，计算 brown 语料库中每个句子的平均词数。

```
C:\Anaconda3>python
Python 3.4.1 |Anaconda 2.1.0 (64-bit)| (default, Sep 24 2014,
18:32:42) [MSC v.1600 64 bit (AMD64)] on win32
Type "help", "copyright", "credits" or "license" for more information.
>>> from nltk.corpus import brown
>>> len(brown.words())/len(brown.sents())
20.250994070456922
from nltk.corpus import brown
sum(len(s) for s in brown.sents())/len(brown.sents())
```

在其他情况下（没有提供 sents 情况下），文本可能只是作为一个字符流。在将文本分词之前，需要将它分割成句子。NLTK 通过包含 Punkt 句子分割器简化了这些操作。

```
>> from nltk import data
>>> sent_tokenizer=data.load('tokenizers/punkt/english.pickle') #模型
>>> text='This is a book. That is a desk.'
>>> sents=sent_tokenizer.tokenize(text) #分句
>>> sents
['This is a book.', 'That is a desk.'] #句点.后必须有空格
>>> [s.split() for s in sents]
[['This', 'is', 'a', 'book.'], ['That', 'is', 'a', 'desk.']]
>>> sum(len(s.split()) for s in sents)/len(sents)
4.0
```

3.8.2 分词

对于一些书写系统，由于没有词边界的可视表示（如中文），文本分词变得更加困难。例如，在中文中，三个字符的字符串爱国人(ai4 "love" [verb], guo3 "country", ren2 "person")，可以被分词为"爱国/人"，"country-loving person"，或者"爱/国人"，"love country-person"。

类似的问题在口语语言处理中也会出现,听者必须将连续的语音流分割成单个的词汇。当我们事先不认识这些词时,这个问题就演变成一个特别具有挑战性的问题。语言学习者会面对这个问题,例如小孩听父母说话。考虑下面人为构造的例子,单词的边界已被去除。

```
a. doyouseethekitty
b. seethedoggy
c. doyoulikethekitty
d. likethedoggy
```

第一个挑战仅仅是表示这个问题,需要找到一种方法来分开文本内容与分词标志。可以给每个字符标注一个布尔值来指示这个字符后面是否有一个分词标志。假设说话人会给语言学习者一个说话时的停顿,这往往是对应一个延长的暂停。这里是一种表示方法,包括初始的分词和最终分词目标。

```
text="doyouseethekittyseethedoggydoyoulikethekittylikethedoggy"
seg1="0000000000000001000000000010000000000000000100000000000"
seg2="0100100100100001001001000010100100010010000100010010000"
```

观察由 0 和 1 组成的分词表示字符串。它们比源文本短一个字符,因为长度为 n 文本可以在 n-1 个地方被分割。函数 segment() 演示了可以从这个表示回到初始分词的文本。

0	1	2	3	4	5	6	7	8	9	10	11	12	13	14	15	16
d	o	y	o	u	s	e	e	t	h	e	k	i	t	t	y	s e
0	0	0	0	0	0	0	0	0	0	0	0	0	0	0	1	0 0
0	1	0	0	1	0	0	0	1	0	0	1	0	0	0	1	0 0

从分词表示字符串 seg1 和 seg2 中重建文本分词。seg1 和 seg2 表示假设的一些儿童讲话的初始和最终分词。函数 segment() 可以使用它们重现分词的文本。

```
def segment(text, segs):
    words=[]
    last=0
    for i in range(len(segs)):
        if segs[i]=='1':
            words.append(text[last:i+1])
            last=i+1
    words.append(text[last:])
    return words

text="doyouseethekittyseethedoggydoyoulikethekittylikethedoggy"
seg1="0000000000000001000000000010000000000000000100000000000"
seg2="0100100100100001001001000010100100010010000100010010000"
segment(text, seg1)
```

```
['doyouseethekitty', 'seethedoggy', 'doyoulikethekitty', 'likethedoggy']
segment(text, seg2)
['do', 'you', 'see', 'the', 'kitty', 'see', 'the', 'doggy', 'do', 'you', 'like', 'the', 'kitty', 'like', 'the', 'doggy']
```

现在分词的任务变成了一个搜索问题，找到将文本字符串正确分割成词汇的字位串。假定学习者接收词，并将它们存储在一个内部词典中。给定一个合适的词典，能够由词典中的词的序列来重构源文本。定义一个目标函数和一个打分函数，使用基于词典的大小和从词典中重构源文本所需的信息量尽力优化它的值。给定一个假设的源文本的分词（左），推导出一个词典和推导表，它能让源文本重构，然后合计每个词项（包括边界标志）与推导表的字符数，作为分词质量的得分。得分值越小表明分词越好。

计算存储词典和重构源文本的成本（LEXICON+DERIVATION）。

```
def segment(text, segs):
    words=[]
    last=0
    for i in range(len(segs)):
        if segs[i]=='1':
            words.append(text[last:i+1])
            last=i+1
    words.append(text[last:])
    return words

def evaluate(text, segs):
    words=segment(text, segs)
    text_size=len(words)
    lexicon_size=len(' '.join(list(set(words))))
    return text_size+lexicon_size

text="doyouseethekittyseethedoggydoyoulikethekittylikethedoggy"
seg1="0000000000000001000000000000001000000000000000000000000"
seg2="0100100100100010010010000101001000100100001000100010000"
seg3="0000100100000011001000000110000100010000001100010000001"
segment(text, seg3)
['doyou', 'see', 'thekitt', 'y', 'see', 'thedogg', 'y', 'doyou', 'like', 'thekit
t', 'y', 'like', 'thedogg', 'y']
evaluate(text, seg3) #46
evaluate(text, seg2) #47
evaluate(text, seg1) #63
```

最后一步是寻找最大化目标函数值 0 和 1 的模式。

使用模拟退火算法的非确定性搜索，一开始仅搜索短语分词。随机扰动 0 和 1，它们与"温度"成比例。每次迭代温度都会降低，扰动边界会减少。

```python
from __future__ import divison
def segment(text, segs):
    words=[]
    last=0
    for i in range(len(segs)):
        if segs[i]=='1':
            words.append(text[last:i+1])
            last=i+1
    words.append(text[last:])
    return words


def evaluate(text, segs):
    words=segment(text, segs)
    text_size=len(words)   #每一个词对应一个字典的索引
    lexicon_size=len(' '.join(list(set(words))))   #词典里的词连接成字符串
    return text_size+lexicon_size
from random import randint
def flip(segs, pos):   #扰动分词表示字符串 segs 的 pos 边界
    return segs[:pos]+str(1-int(segs[pos]))+segs[pos+1:]

def flip_n(segs, n):   #扰动分词表示字符串 n 次
    for i in range(n):
        segs=flip(segs, randint(0, len(segs)-1))
    return segs


def anneal(text, segs, iterations, cooling_rate):
    temperature=float(len(segs))
    while temperature>0.5:
        best_segs, best=segs, evaluate(text, segs)
        for i in range(iterations):   #迭代次数固定
            guess=flip_n(segs, int(round(temperature)))   #开始温度高，扰动次数多
            score=evaluate(text, guess)   #本质上是一个概率算法
            if score<best:   #每次都是在当前最优的基础上扰动迭代，好则更新，不好则丢掉
```

```
            best, best_segs=score, guess
       score, segs=best, best_segs
       temperature=temperature/cooling_rate #退火，扰动次数减少
       print evaluate(text, segs), segment(text, segs)
    print
    return segs

text="doyouseethekittyseethedoggydoyoulikethekittylikethedoggy"
seg1="0000000000000001000000000100000000000000100000000000"
anneal(text, seg1, 5000, 1.2)
 63 ['doyouseethekitty', 'seethedoggy', 'doyoulikethekitty', 'likethedoggy']
 63 ['doyouseethekitty', 'seethedoggy', 'doyoulikethekitty', 'likethedoggy']
 63 ['doyouseethekitty', 'seethedoggy', 'doyoulikethekitty', 'likethedoggy']
 63 ['doyouseethekitty', 'seethedoggy', 'doyoulikethekitty', 'likethedoggy']
 63 ['doyouseethekitty', 'seethedoggy', 'doyoulikethekitty', 'likethedoggy']
 63 ['doyouseethekitty', 'seethedoggy', 'doyoulikethekitty', 'likethedoggy']
 63 ['doyouseethekitty', 'seethedoggy', 'doyoulikethekitty', 'likethedoggy']
 63 ['doyouseethekitty', 'seethedoggy', 'doyoulikethekitty', 'likethedoggy']
 61 ['doyou', 'se', 'ethekitty', 'seet', 'hedoggy', 'doyou', 'lik', 'ethekitty', 'liket', 'hedog', 'gy']
 56 ['doyou', 's', 'e', 'ethekitty', 'se', 'et', 'hedoggy', 'doyou', 'l', 'ik', 'ethekitty', 'lik', 'et', 'hedoggy']
 56 ['doyou', 's', 'e', 'ethekitty', 'se', 'et', 'hedoggy', 'doyou', 'l', 'ik', 'ethekitty', 'lik', 'et', 'hedoggy']
 53 ['doyou', 's', 'e', 'ethekitty', 's', 'e', 'ethedoggy', 'doy', 'ou', 'lik', 'ethekitty', 'lik', 'ethedoggy']
 50 ['doyou', 'se', 'ethekitty', 's', 'e', 'ethedoggy', 'doyou', 'l', 'ik', 'ethekitty', 'l', 'ik', 'ethedoggy']
 48 ['doyou', 'se', 'e', 'thekitty', 'se', 'ethedoggy', 'doyou', 'l', 'ik', 'e', 'thekitty', 'l', 'ik', 'ethedoggy']
 48 ['doyou', 'se', 'e', 'thekitty', 'se', 'ethedoggy', 'doyou', 'l', 'ik', 'e', 'thekitty', 'l', 'ik', 'ethedoggy']
 45 ['doyou', 'se', 'e', 'thekitty', 'se', 'ethedoggy', 'doyou', 'lik', 'e', 'thekitty', 'lik', 'ethedoggy']
 45 ['doyou', 'se', 'e', 'thekitty', 'se', 'ethedoggy', 'doyou', 'lik', 'e', 'thekitty', 'lik', 'ethedoggy']
 42 ['doyou', 'se', 'ethekitty', 'se', 'ethedoggy', 'doyou', 'lik', 'ethekitty', 'lik', 'ethedoggy']
 42 ['doyou', 'se', 'ethekitty', 'se', 'ethedoggy', 'doyou', 'lik',
```

```
'ethekitty', 'lik', 'ethedoggy']
    42 ['doyou', 'se', 'ethekitty', 'se', 'ethedoggy', 'doyou', 'lik',
'ethekitty', 'lik', 'ethedoggy']
    42 ['doyou', 'se', 'ethekitty', 'se', 'ethedoggy', 'doyou', 'lik',
'ethekitty', 'lik', 'ethedoggy']
    42 ['doyou', 'se', 'ethekitty', 'se', 'ethedoggy', 'doyou', 'lik',
'ethekitty', 'lik', 'ethedoggy']
    42 ['doyou', 'se', 'ethekitty', 'se', 'ethedoggy', 'doyou', 'lik',
'ethekitty', 'lik', 'ethedoggy']
    42 ['doyou', 'se', 'ethekitty', 'se', 'ethedoggy', 'doyou', 'lik',
'ethekitty', 'lik', 'ethedoggy']
    42 ['doyou', 'se', 'ethekitty', 'se', 'ethedoggy', 'doyou', 'lik',
'ethekitty', 'lik', 'ethedoggy']
    '0000101000000010100000001000010010000000010010000000'
     d o y o u s e e t h e k i t t y s e e
     0 0 0 0 0 0 0 0 0 0 0 0 0 0 0 1 0 0 0
```

有了足够的数据，就可以以一个合理的准确度自动将文本分割成词汇。这种方法可用于为那些词的边界没有视觉表示的书写系统分词。

3.9 数据制表

格式化字符串的一个重要用途是用于数据制表。对标题和列宽完全控制，要注意语言处理工作与结果制表之间是明确分离的。

brown 语料库不同部分的频率模型。

```
def tabulate(cfdist, words, categories):
    print '%-16s' % 'Category',
    for word in words: #column headings
        print '%6s' % word,
    print
    for category in categories:
        print '%-16s' % category, #row heading
        for word in words: #for each word
            print '%6d' % cfdist[category][word], #print table cell
        print #end the row
```

```
import nltk
from nltk.corpus import brown
cfd=nltk.ConditionalFreqDist(
    (genre, word)
    for genre in brown.categories()
    for word in brown.words(categories=genre))
genres=['news', 'religion', 'hobbies', 'science_fiction', 'romance', 'humor']
modals=['can', 'could', 'may', 'might', 'must', 'will']
tabulate(cfd, modals, genres)

import nltk
from nltk.corpus import brown
cfd=nltk.ConditionalFreqDist(
    (genre, word)
    for genre in brown.categories()
    for word in brown.words(categories=genre))
genres=['news', 'religion', 'hobbies', 'science_fiction', 'romance', 'humor']
modals=['can', 'could', 'may', 'might', 'must', 'will']
cfd.tabulate(samples=modals, conditions=genres)
```

可以使用 width=max(len(w) for w in words)自动定制列的宽度，使其足够容纳所有的词。要记住 print 语句结尾处的逗号增加了一个额外的空格，这样能够防止列标题相互重叠。

小　结

（1）本书将文本作为一个词列表。原始文本是一个潜在的长字符串，其中包含文字和用于设置格式的空白字符，也是通常存储和可视化文本的原料。

（2）在 Python 中指定一个字符串使用单引号、双引号或三引号。

（3）字符串中的字符是使用索引来访问的，索引从零计数。求字符串的长度可以使用 len()。

（4）子字符串使用切片符号来访问。如果省略起始索引，子字符串从字符串的开始处开始；如果省略结尾索引，切片会一直到字符串的结尾处结束。（这点跟 C 语言函数不同）

（5）字符串可以被分割成列表，列表可以连接成字符串。

（6）可以从一个文件 f 读取文本，可以从一个 URL 读取文本，可以遍历一个文本文件的每一行。

（7）在网上找到的文本可能包含不需要的内容，如页眉、页脚和标记，在做语言处理之前需要去除它们。

（8）分词是将文本分割成基本单位或标记，例如词和标点符号等。基于空格符的分词对于许多应用程序都是不够的，因为它会捆绑标点符号和词，NLTK 提供了一个现成的分词器。

（9）词形归并是一个过程，将一个词的各种形式映射到这个词标准的或引用的形式，称为词位或词元。

（10）正则式是用来指定模式的一种强大而灵活的方法，只要导入了 re 模块，就可以使用 re.findall()来找到一个字符串中匹配一个模式的所有子字符串。

（11）当某些字符前使用了反斜杠时，处理时会有特殊的含义；然而，当反斜杠用于正则表达式通配符和操作符时，这些字符失去其特殊的含义，只按字面表示匹配。

练 习

1. 定义一个字符串 s='colorless'，写一个 Python 语句将其变为"colourless"，只使用切片和连接操作。

```
s='colorless'
s=s[:4]+'u'+s[4:]
s
def insert(s, i, e):
    s=s[:i]+e+s[i:]
    return s

s='colorless'
i=s.index('r')
e='u'
s=insert(s, i, e)
s
from re import sub
s='colourless'
sub('r', 'ur', 'colorless')
s
'colorless'.replace('r', 'ur')
```

2. 可以使用切片符号删除词汇形态上的结尾，例如，'dogs'[:-1]删除了 dogs 的最后一个字符，留下 dog。使用切片符号删除下面这些词的词缀，我们插入了一个连字符指示词缀的边界，请在你的字符串中省略掉连字符：dish-es, run-ning, nation-ality, un-do, pre-heat。

```
'dishes'[:-2]
'running'[:-4]
'nationality'[:-5]
'undo'[2:]
'preheat'[3:]
```

3. 我们看到如何产生一个 IndexError，通过索引超出一个字符串的末尾。构造一个向左走得太远走到字符串前面的索引，这有可能吗？

```
s='dishes'
s=s[-80]
'abc'[:-10]
'abc'[-10:]
```

4. 可以为分片指定一个"步长"。下面的表达式间隔一个字符返回一个片内字符：monty[6:11:2]。也可以反向进行：monty[10:5:-2]。自己尝试一下使用不同的步长进行实验。

```
monty='Monty Python'
monty[6:11:2]
monty[10:5:-2]
```

5. 解释器处理 monty[::-1]会发生什么？解释为什么这是一个合理的结果。

```
monty='Monty Python'
monty[::-1]
'nohtyP ytnoM' #reverse
```

6. 说明以下的正则表达式匹配的字符串类。
（1）[a-zA-Z]+；
（2）[A-Z][a-z]*；
（3）p[aeiou]{,2}t；
（4）\d+(\.\d+)?；
（5）([^aeiou][aeiou][^aeiou])*；
（6）\w+|[^\w\s]+。
使用 nltk.re_show()测试你的答案。

```
<字母>→a|b|…|z|A|B|…|Z
L→a|b|…|z|A|B|…|Z
l=a|b|…|z|A|B|…|Z  #e
l [a-zA-Z]  #lex
{a, b, …, z, A, B, …, Z}+
(a|b|…|z|A|B|…|Z)+
```

字母串，字母序列：

```
import nltk
p=r'[a-zA-Z]+'
nltk.re_show(p, '123asd456') #123{asd}456
nltk.re_show(p, '123asd456asd')

p='[A-Z][a-z]*'
nltk.re_show(p, '123asd456asd')
nltk.re_show(p, 'Aadsds123asd456asd')

p='p[aeiou]{,2}t'
nltk.re_show(p, 'paat')
nltk.re_show(p, 'padst')

p='\d+(\.\d+)?'
nltk.re_show(p, '2312.12345dsa') #{2312.12345}dsa
```

7. 写正则表达式匹配下面字符串类。

（1）一个单独的限定符（假设只有 a，an 和 the 为限定符）。

```
import re, nltk
p=r'(a|an|the)' #括号表示范围,匹配字符串用管道符|
re.match(p, 'a an the').group() #'a', a 与 an???贪婪匹配?
re.match(p, 'a an the').span() #(0, 1)
nltk.re_show(p, 'a an the')
```

（2）整数加法和乘法的算术表达式，如 2*3+8。

```
p=r'\d([+*]\d)*'
'''
2*3+8=\d*\d+\d=\di\di\d=\d(i\d){2}=\d(i\d)*=\d([+*]\d)*=r'\d([+*]\d)*'
E→E+E
E→E*E
'''
re.match(p, '1').group()
re.match(p, '1+2').group()
re.match(p, '1+2*3').group()
```

8. 写一个工具函数，以 URL 为参数，返回删除所有 HTML 标记的 URL 的内容。使用 urllib.urlopen 访问 URL 的内容，例如 raw_contents=urllib.urlopen('http://www.nltk.org/').read()。
参见 3.1 节从网络和硬盘访问文本。

```python
import urllib, nltk
raw_contents=urllib.urlopen('http://www.nltk.org/').read()
nltk.clean_html(raw_contents)

from urllib import urlopen
import re
def contents(url):
    raw_contents=urlopen(url).read()
    return re.findall(r'<.*>(.*)<.*>{1, }', raw_contents) #????

url='http://www.nltk.org/'
contents(url)

raw_contents='''
<html>
<head>
<title>Title1</title><title>Title2</title>
</head>
<body>
  Body
</body>
</html>'''
re.findall(r'<.*>(.*)<.*>', raw_contents)
nltk.clean_html(raw_contents)

from urllib import urlopen
import re
def content(url):
    raw_contents=urlopen('http://www.nltk.org/').read()
return re.findall(r'<.*>(.*)<.*>{1, }, raw_contents)
import re
raw_contents='''
<html>
  <head>
    <title>conforming to</title>
  </head>
  <body>
```

```
        consecutive
    </body>
</html>
'''
re.findall(r'<.*>([^<]*)</.*>{1, }', raw_contents)
```

9. 将一些文字保存到文件 corpus.txt。定义一个函数 load(f)，以要读取的文件名为其唯一参数，返回包含文件中文本的字符串。

（1）使用 nltk.regexp_tenize()创建一个分词器分割这个文本中的各种标点符号。使用一个多行的正则表达式，行内要有注释，使用 verbose 标志(?x)。

```
import nltk
def load(file):
    f=open(file)
    return f.read()

text=load('corpus.txt')
print text
pattern=r'''(?x) #set flag to allow verbose regexps
[\]\[., ;"'?():_`\-] #these are separate tens
'''
nltk.regexp_tenize(text, pattern)
[']', '[', '.', ',', ';', '"', "'", '?', '(', ')', ':', '_', '`', '-']
```

（2）使用 nltk.regexp_tenize()创建一个分词器分割以下几种表达式：货币金额；日期；个人和组织的名称。

```
pattern=r'''(?x) #set flag to allow verbose regexps
([A-Z]\.)+ #abbreviations, e.g. U.S.A.
|[A-Z][a-z]*\s[A-Z][a-z]* #words with optional internal
|\d+-\d+-\d+ #注意先后顺序
|\$?\d+(\.\d+)?%? #currency and percentages, e. g. $12.40, 82%
'''
text='U.S.A., $12.40, 82%, 2019-10-11'
nltk.regexp_tenize(text, pattern)
import nltk
text="The bo is $12.34"
pattern=r'''(?x)
\$\d+(\.\d+)?
|[A-Za-z]+'''
```

```
nltk.regexp_tenize(text, pattern)

'''
text.split()
nltk.word_tenize(text)
nltk.wordpunct_tenize(text)
re.split()
'''
import nltk
def load(file):
    f=open(file)
    return f.read()

content=load('corpus.txt')
print content
pattern=r'''(?x)
\w*[., ?:;]'''
nltk.regexp_tenize(content, pattern)
```

10. 将下面的循环改写为列表推导。

```
sent=['The', 'dog', 'gave', 'John', 'the', 'newspaper']
result=[]
for word in sent:
    word_len=(word, len(word))
    result.append(word_len)

result
[('The', 3), ('dog', 3), ('gave', 4), ('John', 4), ('the', 3), ('newspaper', 9)]
```

功能：用单词列表 sent 中的单词及其长度建立一个二元组列表 result。

```
sent=['The', 'dog', 'gave', 'John', 'the', 'newspaper']
result=[(word, len(word)) for word in sent]
result
```

数学语言：

```
result={(word, len(word)) | word ∈ sent}
```

11. 定义一个字符串 raw 包含你自己选择的句子。现在，以空格以外的其他字符分割 raw，如 's'。

```
raw='She sells sea shell by the sea shore .'
sent=raw.split('s')
sent
```

12. 写一个 for 循环输出一个字符串的字符，每行一个。

```
for i in 'Excuse me .':
    print i
```

13. 不带参数的 split 与以' '作为参数的 split，即 sent.split()与 sent.split(' ')的区别是什么？当被分割的字符串包含制表符、连续的空格或一个制表符与空格的序列会发生什么？

```
S.split(sep=None, maxsplit=-1) -> list of strings
Return a list of the words in S, using sep as the delimiter string.
If maxsplit is given, at most maxsplit splits are done. If sep is not
specified or is None, any whitespace string is a separator and empty
strings are removed from the result.
sent=' Asking  for\tnext spot'
sent.split()
sent.split(' ')
['Asking', 'for', 'next', 'spot']
sent=' Asking  for\tnext spot'
sent.split(' ')
['', 'Asking', '', 'for\tnext', 'spot']
import re
sent=' Asking  for\tnext spot'
pattern=r' '
re.split(pattern, sent)
['', 'Asking', '', 'for\tnext', 'spot']
import re
sent=' Asking  for\tnext spot'
pattern=r'\s+'
re.split(pattern, sent)
import re
sent=' Asking  for\tnext spot'
pattern=r'\w+'
re.findall(pattern, sent)
```

14. 创建一个变量 words，包含一个词列表。实验 words.sort()和 sorted(words)，它们有什么区别？

```
words=['The', 'buildings', 'are', 'really', 'magnificent.']
words.sort()
```

list 的方法 sort 是 in place 排序，可以改变自身。

```
words=['The', 'buildings', 'are', 'really', 'magnificent.']
sorted(words)
```

sorted 方法返回排序后的 list，不影响自身。

15. 通过在 Python 提示符输入以下表达式，探索字符串和整数的区别："3"*7 和 3*7。尝试使用 int("3") 和 str(3) 进行字符串和整数之间的转换。

```
"3"*7
3*7
int("3")
str(3)
```

16. 用文本编辑器创建一个名为 test2.py 的文件，只包含一行 monty='Monty Python'。打开一个新的 Python 会话，并在提示符下输入表达式 monty，此时会从解释器得到一个错误。现在，请尝试以下代码（注意要丢弃文件名中的.py）。

```
from test2 import monty
monty
```

Python 会返回一个值。也可以尝试 import test，在这种情况下，Python 应该能够处理提示符处的表达式 test.monty。

```
test2.py, C:\Python27
monty='Monty Python'
monty
from test2 import monty
monty
import test2
test.monty
```

17. 格式化字符串%6s 与%-6s 用来显示长度大于 6 个字符的字符串时，会发生什么？

```
'%6s' % 'Yes . Millions of travellers come here every year for the view of the buildings'
 '%6s' % 'Yes'

'%-6s' % 'Yes . Millions of travellers come here every year for the view of the buildings'
 '%-6s' % 'Yes'
```

18. 阅读语料库中的一些文字，为它们分词，输出其中出现的所有 wh-类型词的列表（英语中的 wh-类型词被用在疑问句、关系从句和感叹句：who，which，what 等）。按顺序输出它们。在这个列表中有因为有大小写或标点符号的存在而重复的词吗？

```
import nltk
from nltk.corpus import brown
pattern=r'''(?x)
wh[a-z]{1, }
|Wh[a-z]{1, }'''
text=nltk.Text(brown.words(categories='news'))
test=str(text[300:1000])
nltk.regexp_tenize(test, pattern)
text=['who', 'which', 'what', 'WHO', 'which', 'what', 'are']
set(sorted(w.lower() for w in text if w.lower().startswith('wh')))
[w for w in set(w.lower() for w in text) if w.startswith('wh')]
```

19. 创建一个文件，包含词汇和（任意指定）频率，其中每行包含一个词、一个空格和一个正整数，如 fuzzy 53。使用 open(filename).readlines() 将文件读入 Python 列表。接下来，使用 split() 将每一行分成两个字段，并使用 int() 将其中的数字转换为一个整数。结果是一个列表形式：[['fuzzy', 53], …]。

```
filename='19.txt' #C:\Python27
[[j[0], int(j[1])] for j in [i.split() for i in open(filename).readlines()]]
result=[]
text=['a 10', 'b 20', 'c 30']
for line in text:
    w, x=line.split()
    result.append([w, int(x)])

result
sent=['fuzzy 53', 'puzzle 63']
sents=[sent[i].split() for i in range(2)]
words=[[sents[i][0], int(sents[i][1])] for i in range(2)]
```

20. 编写代码访问喜爱的网页，并从中提取一些文字。例如，访问一个天气网站，提取你所在的城市今天的最高温度预报。

```
import nltk
from urllib import urlopen
url='http://www.weather.com.cn/weather/101020100.shtml'
```

```
html=urlopen(url).read()
pattern=r'''(?x)'''
nltk.regexp_tenize(html, pattern)
```

21. 写一个函数 unknown()，以一个 URL 为参数，返回一个该网页出现的未知词列表。为了做到这一点，请提取所有由小写字母组成的子字符串（使用 re.findall()），并去除所有在 Words 语料库中出现的项目（nltk.corpus.words）。尝试手动分类这些词，并讨论你的发现。

```
raw='<html>Where are we going to visit today ?</html>'
words=re.findall('[a-z]+', raw.lower())
words=set(words)
dict=['where', 'are', 'we', 'go', 'to', 'visit', 'today']
unknown=[w for w in words if w not in dict]
unknown
```

22. 使用上面建议的正则表达式处理网址 http://news.bbc.co.uk/，检查处理结果。你会看到那里仍然有相当数量的非文本数据，特别是 JavaScript 命令。你可能还会发现句子分割没有被妥善保留。定义更深入的正则表达式，改善此网页文本的提取。

23. 你能写一个正则表达式以这样的方式来分词吗？将词 don't 分为 do 和 n't？解释为什么这个正则表达式无法正常工作：«n't|\w+»。

```
from re import *
p=r"n't|\w+"
word="don't"
findall(p, word)
from re import *
p=r"do|n't"
word="don't"
findall(p, word)
from re import *
p=r"t'n|\w+"
word="t'nod"
findall(p, word)
```

24. 尝试编写代码将文本转换成 hAck3r，使用正则表达式和替换，其中 e→3，i→1，o→0，l→|，s→5，.→5w33t!，ate→8。在转换之前将文本规范化为小写。自己添加更多的替换。现在尝试将 s 映射到两个不同的值：词开头的 s 映射为$，词内部的 s 映射为 5。

```
from re import *
text=['We', 'will', 'visit', 'the', 'Statue', 'of', 'Liberty', 'as', 'scheduled', '.']
a='^s'
```

```
b='$'
c='s'
d='5'
[sub(c, d, sub(a, b, word.lower())) for word in text]
import re
text=['We', 'will', 'visit', 'the', 'Statue', 'of', 'Liberty', 'as', 'scheduled', '.']
text=[re.sub('^s', '$', word.lower()) for word in text]
[re.sub('s', '5', word) for word in text]

re.sub(r'^s', '$', 'We will visit the Statue of Liberty as scheduled.'.lower())
re.sub(r's', '5', 'We will visit the Statue of Liberty as scheduled.'.lower())

p1=r'e'
p2=r'i'
p3='o'
p4=r'[.]'
p5=r'ate'
p6=r'^s'
p7=r's'
p8=r'1'

def f(s):
    global s1
    s=re.sub(p1, '3', s)
    s=re.sub(p2, '1', s)
    s=re.sub(p3, '0', s)
    s=re.sub(p4, '5w33t!', s)
    s=re.sub(p5, '8', s)
    s=re.sub(p6, '$', s)
    s=re.sub(p7, '5', s)
    s=re.sub(p8, '|', s)
    s1=s
s=re.sub(r'e', '3', 'We will visit the Statue of Liberty as scheduled.')
s1='eio'
f(s1.lower())
```

25. Pig Latin 是英语文本的一个简单的变换。文本中每个词的按如下方式变换：将出现在词首的所有辅音（或辅音群）移到词尾，然后添加 ay，例如，string→ingstray, idle→idleay。

（1）写一个函数转换一个词为 Pig Latin。

（2）写代码转换文本而不是单个的词。

（3）进一步扩展它，保留大写字母，将 qu 保持在一起（例如，这样 quiet 会变成 ietquay），并检测 y 是作为一个辅音（如 yellow）还是元音（如 style）。

```python
def Pig_Latin(s):
    s=s.lower()  #转换为小写
    #以元音字母开始，则在单词末尾加入"hay"
    if s[0] in 'aeiou':
        vowel_begin=True
    else:
        vowel_begin=False
    if vowel_begin:
        s=s+'hay'
    #以'q'字母开始，并且后面有个字母'u'，将"qu"移动到单词末尾加入"ay"
    if s[0]=='q' and s[1]=='u':
        s =s[2:]+'quay'
    #以辅音字母开始，所有连续的辅音字母一起移动到单词末尾加入"ay"
    end=0
    for i in range(len(s)):
        if s[i] in 'aeiou' or (i>0 and s[i]=='y'):
            end=i
            break
    s=s[end:]+s[:end]+'ay'
    return s

def transfer_main(str):
    s=str.split()
    result=[]
    for i in s:
        result.append(Pig_Latin(i))
    new_s=' '.join(result)
    return new_s

print transfer_main('Welcome to the Python world Are you ready')
```

26. 下载一种包含元音和谐语言（如匈牙利语）的一些文本，提取词汇的元音序列，并创建一个元音二元语法表。

元音和谐是同化的一种特殊类型，如土耳其语"ip-ler"（线，复数）和"baba-lar"（父亲，复数）中的"ler""ar"都表示复数，但元音不同，这和它们前面的元音有关。

27. Python 的 random 模块包括函数 choice()，它从一个序列中随机选择一个项目。例如，choice("aehh")会产生四种可能的字符中的一个，字母 h 的概率是其他字母的两倍。写一个表达式产生器，从字符串"aehh"产生 500 个随机选择的字母序列，并将这个表达式写入函数".join()调用中，将它们连接成一个长字符串。你得到的结果应该看起来像失去控制的喷嚏或狂笑：he haha ee heheeh eha。使用 split()和 join()再次规范化这个字符串中的空格。

```
from random import *
' '.join(''.join(choice("aehh ") for i in range(50)).split())
```

28. 考虑下面摘自 MedLine 语料库句子中的数字表达式：The corresponding free cortisol fractions in these sera were 4.53+/-0.15% and 8.16+/-0.23%, respectively。我们应该说数字表达式 4.53 +/- 0.15%是三个词吗？或者我们应该说它是一个单独的复合词？它实际上是 9 个词，因为它读作"four point five three, plus or minus fifteen percent"，或，说这不是一个"真正的"词，因为它不会出现在任何词典中。讨论这些不同的可能性。你能想出产生这些答案中两种以上可能性的应用领域吗？

29. 可读性测量用于为一个文本的阅读难度打分，给语言学习者挑选适当难度的文本。在一个给定的文本中，定义 μ_w 为每个词的平均字母数，μ_s 为每个句子的平均词数。文本自动可读性指数（Automated Readability Index，ARI）被定义为：4.71μ_w+0.5μ_s-21.43。计算 brown 语料库各部分的 ARI 得分，包括 f 部分 popular(lore)和 j(learned)。利用 nltk.corpus.brown.words()产生一个词汇序列，利用 nltk.corpus.brown.sents()产生一个句子的序列。

```
import nltk
from __future__ import division
words=nltk.corpus.brown.words(categories='learned')
sents=nltk.corpus.brown.sents(categories='learned')
uw=sum([len(w) for w in words])/len(words)
us=len(words)/len(sents)
ARI=4.71*uw+0.5*us-21.43
ARI    #11.926007043317348
import nltk
from __future__ import division
text='1234 12345 . 123 1234 ?'
words=nltk.word_tenize(text)
sents=nltk.sent_tenize(text)
uw=sum([len(w) for w in words])/len(words)
```

```
us=len(words)/len(sents)
ARI=4.71*uw+0.5*us-21.43
ARI
-5.800000000000001
import nltk
from __future__ import division
words=nltk.corpus.brown.words(categories='lore')
sents=nltk.corpus.brown.sents(categories='lore')
uw=sum([len(w) for w in words])/len(words)
us=len(words)/len(sents)
ARI=4.71*uw+0.5*us-21.43
ARI #10.254756197101155
```

learned 的 ARI=11.926007043317348>lore 的 ARI=10.254756197101155，所以 learned 比 lore 要难些。

30. 使用 Porter 词干提取器规范化一些已标注的文本的过程叫作为每个词提取词干。用 Lancaster 词干提取器做同样的事情，看看你是否能观察到一些区别。

31. 定义变量 saying 包含列表['After', 'all', 'is', 'said', 'and', 'done', ',', 'more', 'is', 'said', 'than', 'done', '.']。使用 for 循环处理这个列表，并将结果存储在一个新的列表 lengths 中。提示：使用 lengths=[]，分配一个空列表给 lengths。然后每次循环中用 append()添加另一个长度值到列表中。

```
saying=['After', 'all', 'is', 'said', 'and', 'done', ',', 'more',
'is', 'said', 'than', 'done', '.']
lengths=[]
for w in saying:
    lengths.append(w)

lengths=[w for w in saying]
```

32. 定义一个变量 silly 包含字符串：'newly formed bland ideas are inexpressible in an infuriating way'。现在编写代码执行以下任务。

（1）分割 silly 为一个字符串列表，每一个词一个字符串，使用 Python 的 split()操作，将结果保存到叫作 bland 的变量中。

（2）提取 silly 中每个词的第二个字母，将它们连接成一个字符串，得到'eoldrnnnna'。

（3）使用 join()将 bland 中的词组合回一个单独的字符串。确保结果字符串中的词以空格隔开。

（4）按字母顺序输出 silly 中的词，每行一个。

```
silly='newly formed bland ideas are inexpressible in an infuriating way'
bland=silly.split()
```

```
#from functools import reduce
s=reduce(lambda x, y:x+y, [w[1] for w in bland])
' '.join(bland)
sorted(bland)


silly='newly formed bland ideas are inexpressible in an infuriating way'
bland=silly.split()
s=''.join([w[1] for w in bland])
' '.join(bland)
sorted(bland)
```

33. index()函数可用于查找序列中的项目。例如，'inexpressible'.index('e')告诉我们字母 e 的第一个位置的索引值。

（1）当你查找一个子字符串如'inexpressible'.index('re')时会发生什么？

```
'inexpressible'.index('e')  #2
'inexpressible'.index('re') #5
```

（2）定义一个变量 words 包含一个词列表。现在使用 words.index()来查找一个单独的词的位置。

```
words=['The', 'journey', 'is', 'worth', 'while', '.', 'Where',
'are', 'we', 'going', 'after', 'this', '?']
words.index('The') #0
```

（3）定义一个变量 silly，使用 index()函数结合列表切片，建立一个包括 silly 中 in 之前（但不包括 in）的所有的词的列表 phrase。

```
silly='newly formed bland ideas are inexpressible in an infuriating way'
p=silly.index(' in ')
silly[:p]
```

34. 编写代码，将国家的形容词转换为它们对应的名词形式，如将 Canadian 和 Australian 转换为 Canada 和 Australia。

35. 阅读 LanguageLog 中关于短语的 as best as p can 和 as best p can 形式的帖子，其中 p 是一个代名词。在一个语料库和 3.5 节中描述搜索已标注（已分词）文本 findall()方法的帮助下，调查这一现象。

```
import nltk
raw=open('Taming.txt').read()
words=nltk.word_tenize(raw)
hobbies_learned=nltk.Text(words)
hobbies_learned.findall(r"<as><best><as><.*><can>")
```

```
hobbies_learned.findall(r"<as><best><.*><can>")
as best <SUBJ> <MODAL>
```

36. 阅读有关 re.sub()函数使用正则表达式进行字符串替换的内容。使用 re.sub 编写代码从一个 HTML 文件中删除 HTML 标记，规范化空格。

```
import re
html='''
<html>
  <head>
    <title>conforming to<title>
  </head>
  <body>
    consecutive
  </body>
</html>
'''
p=re.compile(r'<[^>]+>')
content=p.sub('', html)
#content=re.sub(' +', ' ', content)
' '.join(re.findall('\w+', content))
```

37. 分词的一个有趣的挑战是已经被分割的跨行词，例如，如果 long-term 被分割，我们能得到字符串 long-\nterm。

（1）写一个正则表达式，识别连字符连结的跨行处词汇。这个表达式将需要包含\n 字符。

（2）使用 re.sub()从这些词中删除\n 字符。

（3）你如何确定一旦换行符被删除后不应该保留连字符的词汇，如'encyclo-\npedia'。

```
import re
line='duplicate consecutive tencyclo-\npedia conforming to'
line=re.findall(r'[\S\n]+', line)
[re.sub(r'-\n', '', w) for w in line]


line='duplicate consecutive encyclo-\npedia conforming to'
line.replace('-\n', '')

#line=nltk.word_tenize(line)
re.sub(r'-\n', '', line)
```

38. Soundex 算法是一种拼音算法，用于按英语发音来索引姓名。Soundex 方法返回一个表示姓名的四字符代码，由一个英文字母后跟三个数字组成。字母是姓名的首字母，数字对姓名中剩余的辅音字母编码。发音相近的姓名具有相同的 Soundex 代码。辅音字母将映射到一个特定的数字。

(1)将英文字按以下规则替换(除第一个字符外)。

```
a e h i o u w y → 0
b f p v → 1
c g j k q s x z → 2
d t → 3
l → 4
m n → 5
r → 6
```

(2)去除 0,对于重复的字符只保留一个。
(3)返回前 4 个字符,不足 4 位以 0 补足。

```
""" a e h i o u w y → 0
    b f p v → 1
    c g j k q s x z → 2
    d t → 3
    l → 4
    m n → 5
    r → 6
"""
#        '012345678910 11 12 13 14 15 16 17 18 19 20 21 22 23 24 25'
#        'abcdefghijklmnopqrstuvwxyz'
# digits='01230120022455012623010202'

def soundex(name, len=4):
    """ soundex module conforming to Knuth's algorithm
        implementation 2000-12-24 by Gregory Jorgensen
        public domain
    """
    digits='01230120022455012623010202' #digits holds the soundex values for the alphabet
    sndx=''
    fc=''
    #translate alpha chars in name to soundex digits
    for c in name.upper():
        if c.isalpha():
            if not fc:
                fc=c #remember first letter
            d=digits[ord(c) - ord('A')]
```

```
            if not sndx or (d != sndx[-1]): #duplicate consecutive soundex digits are skipped
                sndx += d
        print sndx
        sndx=fc+sndx[1:] #replace first digit with first alpha character
        sndx=sndx.replace('0', '') #remove all 0s from the soundex code
        return (sndx+(len*'0'))[:len] #return soundex code padded to len characters

    soundex('kant')
    #2053
    #'K530'
    soundex('Knuth')
    #25030
    #'K530'
```

39. 获取两个或多个文体的原始文本，计算它们各自的在前面关于阅读难度的练习中描述的阅读难度得分。例如，比较 ABC 农村新闻和 ABC 科学新闻（nltk.corpus.abc）。使用 Punkt 处理句子分割。

```
import nltk
from __future__ import division
from nltk.corpus import abc
words=abc.words('rural.txt')
sents=abc.sents('rural.txt')
uw=sum(len(w) for w in words)/len(words)
us=len(words)/len(sents)
ARI=4.71*uw+0.5*us-21.43
ARI #12.147911045270774
from nltk.corpus import abc
abc.fileids()
['rural.txt', 'science.txt']
import nltk
from __future__ import division
from nltk.corpus import abc
words=abc.words('science.txt')
sents=abc.sents('science.txt')
uw=sum(len(w) for w in words)/len(words)
us=len(words)/len(sents)
```

```
ARI=4.71*uw+0.5*us-21.43
ARI #12.409976957108434
ABC 科学新闻比 ABC 农村新闻要难些。
import nltk
from __future__ import division
from nltk.corpus import abc
from nltk import data
words=abc.words('rural.txt')
sent_tenizer=data.load('tenizers/punkt/english.pickle')
text=abc.raw('rural.txt')
sents=sent_tenizer.tenize(text)
uw=sum(len(w) for w in words)/len(words)
us=len(words)/len(sents)
ARI=4.71*uw+0.5*us-21.43
ARI #12.365180041515053
```

40. 将下面的嵌套循环重写为嵌套列表推导。

```
words=['attribution', 'confabulation', 'elocution', 'sequoia',
'tenacious', 'unidirectional']
vsequences=set()
for word in words:
    vowels=[]
    for char in word:
        if char in 'aeiou':
            vowels.append(char)
    vsequences.add(''.join(vowels))

sorted(vsequences)
#['aiuio', 'eaiou', 'eouio', 'euoia', 'oauaio', 'uiieioa']
words=['attribution', 'confabulation', 'elocution', 'sequoia',
'tenacious', 'unidirectional']
sorted(''.join(char for char in word if char in 'aeiou') for word in words)
```

41. 使用 WordNet 为一个文本集合创建语义索引。扩展一致性搜索程序，使用它的第一个同义词集偏移索引每个词，如 wn.synsets('dog')[0].offset（或者使用上位词层次中的一些祖先的偏移，这是可选的）。

```
from nltk.corpus import wordnet as wn
wn.synsets('dog')[0].offset
```

使用词干提取器索引文本。

```
import nltk
class IndexedText(object):
    def __init__(self, stemmer, text):
        self._text=text
        self._stemmer=stemmer
        self._index=nltk.Index((self._stem(word), i) for i, word in enumerate(text))
    def concordance(self, word, width=40): #不是有内置的吗？
        key=self._stem(word)
        wc=width/4 # words of context
        for i in self._index[key]:
            lcontext=' '.join(self._text[i-wc:i])
            rcontext=' '.join(self._text[i:i+wc])
            ldisplay='%*s' % (width, lcontext[-width:]) #*,输出宽度参数!!!
            rdisplay='%-*s' % (width, rcontext[:width]) #-,左对齐!!!
            print ldisplay, rdisplay
    def _stem(self, word):
        return self._stemmer.stem(word).lower()

porter=nltk.PorterStemmer()
grail=nltk.corpus.webtext.words('grail.txt') #grail 圣杯
text=IndexedText(porter, grail)
text.concordance('lie')
```

42. 在多语言语料库、NLTK 的频率分布和关系排序的功能（nltk.FreqDist，nltk.spearman_correlation）的帮助下，开发一个系统，猜测未知文本。为简单起见，使用一个单一的字符编码和几种语言。

在统计学中，以查尔斯·斯皮尔曼命名的斯皮尔曼等级相关系数，即 spearman 相关系数，经常用希腊字母 ρ 表示。它是衡量两个变量的依赖性的非参数指标。它利用单调方程评价两个统计变量的相关性。

43. 写一个程序处理文本，发现一个词以一种新的意义被使用的情况。对于每一个词计算这个词所有同义词集与这个词的上下文的所有同义词集之间的 WordNet 相似性（注意，这是一个粗略的办法，要做得很好是很困难的，此问题属于开放性研究问题）。

44. 阅读关于规范化非标准词的文章，实现一个类似的文字规范系统。

阅读材料

1. TF-IDF 实现文本分类

TF-IDF 是 Term Frequency - Inverse Document Frequency 的缩写，即词频-逆文本频率。它由两部分组成，TF 和 IDF。前面的 TF 也就是我们前面说到的词频，我们之前做的向量化也就是做了文本中各个词的出现频率统计，并作为文本特征，这个很好理解。关键是后面的 IDF，即逆文本频率如何理解。几乎所有文本都会出现的"to"，其词频虽然高，但是重要性却应该比词频低的"China"和"Travel"要低。IDF 是反映这个词的重要性的，进而修正仅仅用词频表示的词特征值。概括来讲，IDF 反映了一个词在所有文本中出现的频率，如果一个词在很多的文本中出现，那么它的 IDF 值应该低，如上文中的"to"。而反过来如果一个词在比较少的文本中出现，那么它的 IDF 值应该高。例如一些专业的名词，如"Machine Learning"，这样的词 IDF 值应该高。一个极端的情况，如果一个词在所有的文本中都出现，那么它的 IDF 值应该为 0。

上面是定性说明 IDF 的作用，那么如何对一个词的 IDF 进行定量分析呢？这里直接给出一个词 x 的 IDF 的基本公式：

$$\text{IDF}(x) = \log \frac{N}{N(x)}$$

式中，N 代表语料库中文本的总数；$N(x)$ 代表语料库中包含词 x 的文本总数。为什么 IDF 的基本公式应该是上面这样的，而不是像 $N/N(x)$ 这样的形式呢？这就涉及信息论相关的一些知识了，感兴趣的读者建议阅读吴军博士的《数学之美》第 11 章。

上面的 IDF 公式在一些特殊的情况会有一些小问题，比如某一个生僻词在语料库中没有，分母为 0，IDF 便没有意义了。所以，常用的 IDF 需要做一些平滑处理，使语料库中没有出现的词也可以得到一个合适的 IDF 值，平滑的方法有很多种，最常见的 IDF 平滑后的公式之一为

$$\text{IDF}(x) = \log \frac{N+1}{N(x)+1} + 1$$

有了 IDF 的定义，就可以计算某一个词的 TF-IDF 值：

$$\text{TF-IDF}(x) = \text{TF}(x) * \text{IDF}(x)$$

式中，TF(x) 指词 x 在当前文本中的词频。代码实现文本分类如下：

```
import os
import jieba
from sklearn.feature_extraction.text import TfidfTransformer
from sklearn.feature_extraction.text import CountVectorizer
from sklearn.naive_bayes import MultinomialNB
# 以空格来划分每一个分词
```

```python
    def preprocess(path):
        text_with_space = ""
        textfile = open(path, "r", encoding="utf-8").read()
        textcute = jieba.cut(textfile)
        for word in textcute:
            text_with_space += word + " "
            # print(text_with_space)
        return text_with_space

    def loadtrainset(path, classtag):
        allfiles = os.listdir(path)  # os.path.isdir()用于判断对象是否为一个目录,并返回此目录下的所有文件名
        processed_textset = []
        allclasstags = []
        for thisfile in allfiles:
            # print(thisfile)
            path_name = path + "/" + thisfile
            processed_textset.append(preprocess(path_name))
            allclasstags.append(classtag)
        return processed_textset, allclasstags  # 数组形式--processed_textset 文件的具体内容, allclasstags 文件分类

    processed_textdata1, class1 = loadtrainset("C:/Desk/MyProjects/Python/NLP/dataset/train/hotel", "宾馆")
    processed_textdata2, class2 = loadtrainset("C:/Desk/MyProjects/Python/NLP/dataset/train/travel", "旅游")
    train_data = processed_textdata1 + processed_textdata2
    # print(train_data)  # 前半部分是宾馆,后半部分是旅游 train
    classtags_list = class1 + class2  # 前半部分是宾馆,后半部分是旅游 train 集结果
    # print(train_data)
    # print(classtags_list)
    """
    # CountVectorizer是通过fit_transform函数将文本中的词语转换为词频矩阵
        get_feature_names()可看到所有文本的关键字
        vocabulary_可看到所有文本的关键字和其位置
        toarray()可看到词频矩阵的结果
    """
    count_vector = CountVectorizer()
```

```python
    vecot_matrix = count_vector.fit_transform(train_data)
    # print (count_vector.get_feature_names ())    #看到所有文本的关键字
    # print (count_vector.vocabulary_)     #文本的关键字及其位置
    # print (vecot_matrix.toarray ())   #词频矩阵的结果
    # #TFIDF
    """TfidfTransformer 是统计 CountVectorizer 中每个词语的 tf-idf 权值
    tfidf                                                                    =
transformer.fit_transform(vectorizer.fit_transform(corpus))
    vectorizer.fit_transform(corpus)将文本 corpus 输入，得到词频矩阵
    将这个矩阵作为输入，用 transformer.fit_transform(词频矩阵)得到 TF-IDF
权重矩阵
    TfidfTransformer + CountVectorizer  =  TfidfVectorizer
    这个成员的意义是词典索引，对应的是 TF-IDF 权重矩阵的列，只不过一个是私有成
员，一个是外部输入，原则上应该保持一致
       use_idf:boolean, optional 启动 inverse-document-frequency 重新计
算权重
    """
    # print(train_tfidf)  # vecot_matrix 输入，得到词频矩阵
    train_tfidf = TfidfTransformer(use_idf=False).fit_transform(vecot_matrix)
    #MultinomialNB(), fit() ，多分类，Fit Naive Bayes classifier
according to X, 根据 X, Y 结果类别，进行多分类
    # print(train_tfidf)
    # print(classtags_list)
    clf = MultinomialNB().fit(train_tfidf, classtags_list)
    testset = []
    path = "C:/Desk/MyProjects/Python/NLP/dataset/tt"
    allfiles = os.listdir(path)
    hotel = 0
    travel = 0
    for thisfile in allfiles:
        path_name = path + "/" + thisfile  # 得到此目录下的文件绝对路径
        new_count_vector = count_vector.transform([preprocess(path_name)])
# 得到测试集的词频矩阵
    # 用 transformer.fit_transform(词频矩阵)得到 TF-IDF 权重矩阵
        new_tfidf = TfidfTransformer(use_idf=False).fit_transform(new_count_vector)
        # 根据由训练集而得到的分类模型 clf，由测试集的 TF-IDF 权重矩阵来进行预
测分类
        predict_result = clf.predict(new_tfidf)
```

```
        print(predict_result)
        print(thisfile)
        if(predict_result == "宾馆"):
            hotel += 1
        if(predict_result == "旅游"):
            travel += 1

print("宾馆" + str(hotel))
print("旅游" + str(travel))
```

2. Python 2.7.11 下中文文件读写及打印

读写中文文件时，不需要考虑编码的情况，此时虽然可以正常从文件中读取中文，也可以正常地将中文写入文件中，但是无法正常打印中文字段到屏幕上。

src.txt，编码 UTF-8：

```
北京。
上海。
Hello world!
Hello python!
>>> SRC_PATH='src.txt'
>>> DST_PATH='dst.txt'
>>> src_file=open(SRC_PATH, 'r')
>>> dst_file=open(DST_PATH, 'w')
>>> for line in src_file.readlines():
...     dst_file.writelines(line)
...     print line.strip()
...
>>> src_file.close()
>>> dst_file.close()
```

打印：

```
锘胯    鏉夸腑寒寫樅鎝榆含鉑
閲戣瀺涓    繚鏂尢笓娴枫€
Hello world!
Hello python!
```

dst.txt，编码 UTF-8：

```
北京。
上海。
Hello world!
Hello python!
```

打印中文字段时，需要提前把系统编码由 ASCII 转换到 UTF-8：

```
>>> SRC_PATH='src.txt'
>>> DST_PATH='dst.txt'
>>> import sys
>>> reload(sys)
<module 'sys' (built-in)>
>>> sys.setdefaultencoding('utf-8')
>>> src_file=open(SRC_PATH, 'r')
>>> dst_file=open(DST_PATH, 'w')
>>> for line in src_file.readlines():
...     dst_file.writelines(line)
...     print line.strip().encode('gb18030')
...
北京。
上海。
Hello world!
Hello python!
>>> src_file.close()
>>> dst_file.close()
```

查看系统编码的具体转换状况：

```
import sys
print('origin_encoding={}'.format(sys.getdefaultencoding()))
reload(sys)
sys.setdefaultencoding('utf-8')
print ('new_encoding={}'.format(sys.getdefaultencoding()))
```

在不转换系统编码下直接输出中文字段：

```
print u'中文'
print u'中文'.encode('gbk')
print u'中文'.encode('gb18030')
print
print '中文'
print u'中文'.encode('utf-8')
```

在转换系统编码下直接输出中文字段：

```
import sys
reload(sys)
sys.setdefaultencoding('utf-8')
```

```
print u'中文'
print '中文'.encode('gbk')
print '中文'.encode('gb18030')
print u'中文'.encode('gbk')
print u'中文'.encode('gb18030')
print
print '中文'
print '中文'.encode('utf-8')
print u'中文'.encode('utf-8')
```

第4章 结构化程序设计

4.1 序 列

4.1.1 序列的分割

对于一些 NLP 的任务，需要用 90%的数据来训练一个系统，剩余 10%进行测试，要做到这一点，只要指定想要分割数据的位置，然后在这个位置分割序列即可。

```
from nltk.corpus import nps_chat
text=nps_chat.words()
id(text)
cut=int(0.9*len(text))
training_data, test_data=text[:cut], text[cut:]
text == training_data+test_data
#此过程中原始数据没有丢失，也不是复制
id(text)
id(training_data)
text is training_data+test_data
len(training_data)/len(test_data)   #两块大小的比例是预期的

F:\Anaconda>python
Python 2.7.9 |Anaconda 2.2.0 (64-bit)| (default, Dec 18 2014, 16:57:52)
[MSC v.1500 64 bit (AMD64)] on win32
Type "help", "copyright", "credits" or "license" for more information.
Anaconda is brought to you by Continuum Analytics.
Please check out: http://continuum.io/thanks and https://binstar.org
>>> from nltk.corpus import nps_chat
>>> text=nps_chat.words()
>>> id(text)
333023720L
>>> cut=int(0.9*len(text))
>>> training_data, test_data=text[:cut], text[cut:]
```

```
>>> text == training_data+test_data
True
>>> #此过程中原始数据没有丢失，也不是复制
... id(text)
333023720L
>>> id(training_data)
333122920L
>>> text is training_data+test_data
False
>>> len(training_data)/len(test_data)  #两块大小的比例是预期的
9
```

4.1.2 序列的嵌套

一个列表是一个典型的具有相同类型对象的序列，它的长度是任意的（类似于数据结构的线性表），我们经常使用列表保存词序列。相反，一个元组通常是不同类型的对象的集合，长度固定（类似于 C 语言的结构体），我们经常使用一个元组来保存一个记录与一些实体相关的不同字段的集合。

在 Python 中，列表是可变的，而元组是不可变的。换句话说，列表可以被修改，而元组不能。这里是一些在列表上的操作，例如修改一个列表：

```
>>> lexicon=[('the', 'det', ['Di:', 'D@']), ('off', 'prep', ['Qf', 'O:f'])]
>>> lexicon
[('the', 'det', ['Di:', 'D@']), ('off', 'prep', ['Qf', 'O:f'])]
>>> lexicon.sort()
>>> lexicon
[('off', 'prep', ['Qf', 'O:f']), ('the', 'det', ['Di:', 'D@'])]
>>> lexicon[1]=('turned', 'VBD', ['t3:nd', 't3`nd'])
>>> lexicon
[('off', 'prep', ['Qf', 'O:f']), ('turned', 'VBD', ['t3:nd', 't3`nd'])]
>>> del lexicon[0]
>>> lexicon
[('turned', 'VBD', ['t3:nd', 't3`nd'])]
```

使用 lexicon=tuple(lexicon)将词典转换为一个元组，然后尝试上述操作，确认它们都不能运用在元组上。

```
>>> lexicon=[('the', 'det', ['Di:', 'D@']), ('off', 'prep', ['Qf', 'O:f'])]
>>> lexicon=tuple(lexicon)
>>> lexicon
```

```
(('the', 'det', ['Di:', 'D@']), ('off', 'prep', ['Qf', 'O:f']))
>>> lexicon.sort()
Traceback (most recent call last):
  File "<stdin>", line 1, in <module>
AttributeError: 'tuple' object has no attribute 'sort'
>>> lexicon
(('the', 'det', ['Di:', 'D@']), ('off', 'prep', ['Qf', 'O:f']))
>>> lexicon[1]=('turned', 'VBD', ['t3:nd', 't3`nd'])
Traceback (most recent call last):
  File "<stdin>", line 1, in <module>
TypeError: 'tuple' object does not support item assignment
>>> lexicon
(('the', 'det', ['Di:', 'D@']), ('off', 'prep', ['Qf', 'O:f']))
>>> del lexicon[0]
Traceback (most recent call last):
  File "<stdin>", line 1, in <module>
TypeError: 'tuple' object doesn't support item deletion
>>> lexicon
(('the', 'det', ['Di:', 'D@']), ('off', 'prep', ['Qf', 'O:f']))
```

4.2 风　格

4.2.1 过程风格与声明风格

以不同的方式执行相同的任务，蕴含着对执行效率的影响，另一个影响程序开发的是编程风格。下面是计算 brown 语料库中词平均长度的程序。

```
>>> from nltk.corpus import brown
>>> tokens=brown.words(categories='news')
>>> count=0
>>> total=0
>>> for token in tokens:
...     count+=1
...     total+=len(token)
...
>>> print(total/count)
4.401545438271973
>>> tokens='This is a book .'.split()
```

```
>>> count=0
>>> total=0
>>> for token in tokens:
...     count+=1
...     total+=len(token)
...
>>> print(total/count)
2.4
>>> tokens
['This', 'is', 'a', 'book', '.']
>>> count  #单词数
5
>>> total  #字符数
12
```

在这段程序中，使用变量 count 跟踪遇到的标识符的数量，total 储存所有词长度的总和，这是一个低级别的风格，与机器代码，即计算机的 CPU 所执行的基本操作相差不远。两个变量就像 CPU 的两个寄存器，积累许多中间环节产生的值，故可以说这段程序是以过程风格编写（面向过程）。

```
>>> tokens='This is a book .'.split()
>>> total=sum(len(t) for t in tokens)
>>> print(total/len(tokens))
2.4
```

使用生成器表达式累加标识符的长度，每行代码执行一个完整的、有意义的工作，可以高级别的方式来理解，实施细节留给 Python 解释器。第二段程序使用内置函数，在一个更抽象的层面构成程序，生成的代码可读性更好。

下面看一个极端的例子：

```
>>> tokens=['b', 'a', 'b']  #tokens 去重排序到 word_list
>>> word_list=[]
>>> len_word_list=len(word_list)
>>> i=0
>>> while i<len(tokens):
...     j=0
...     while j<len_word_list and word_list[j]<tokens[i]:
...         j+=1  #定位
...     if j==0 or tokens[i]!=word_list[j]:  #去重
...         word_list.insert(j, tokens[i])  #线性表=>有序表
...         len_word_list+=1  #表长加 1
```

```
...     i+=1
...
>>> tokens
['b', 'a', 'b']
>>> word_list
['a', 'b']
```

等效的声明版本使用熟悉的内置函数，可以立即知道代码的功能，word_list= sorted(set(tokens))。

另一种情况，对于每行输出一个计数值，一个循环计数器似乎是必要的。然而，我们可以使用 enumerate()处理序列 s，为 s 中每个项目产生一个(i，s[i])形式的元组，以(0，s[0])开始。下面枚举频率分布的值，捕获变量 rank 和 word 中的整数-字符串对，按照产生排序项列表时的需要，输出 rank+1 使计数从 1 开始。

```
>>> from nltk import FreqDist
>>> from nltk.corpus import brown
>>> fd=FreqDist("This is a book . That is a desk .".split())
>>> cumulative=0.0
>>> "This is a book . That is a desk .".split()
['This', 'is', 'a', 'book', '.', 'That', 'is', 'a', 'desk', '.']
>>> fd
FreqDist({'a': 2, 'is': 2, '.': 2, 'That': 1, 'This': 1, 'desk': 1, 'book': 1})
>>> fd.N()
10
>>> list(enumerate(fd))
[(0, 'a'), (1, 'is'), (2, 'That'), (3, 'This'), (4, 'desk'), (5, '.'), (6, 'book')]
>>> for rank, word in enumerate(fd):
...     cumulative+=fd[word]*100/fd.N()
...     print("%3d %6.2f%% %s" % (rank+1, cumulative, word))
...
  1  20.00% a
  2  40.00% is
  3  50.00% That，其实并没有 rank
  4  60.00% This
  5  70.00% desk
  6  90.00% .
  7 100.00% book
```

使用循环变量存储最大值或最小值，查找出文本中最长的词。

```
>>> text="This is a book . That is a desk .".split()
>>> longest=''
>>> for word in text:
...     if len(word)>len(longest):
...         longest=word
...
>>> longest
'This'
```

另一个解决方案是使用两个列表推导。

```
>>> maxlen=max(len(word) for word in text)
>>> [word for word in text if len(word)==maxlen]
['This', 'book', 'That', 'desk']
```

注意：第一个解决方案找到第一个长度最长的词（最优解），而第二种方案找到所有最长的词（通常是我们想要的所有解）。两个解决方案的理论效率存在差异，主要的开销是到内存中读取数据，一旦数据准备好，第二阶段处理数据可以瞬间高效完成。

4.2.2 循环变量的使用

在某些情况下，我们仍然要在列表推导中使用循环变量，例如，需要使用一个循环变量提取列表中连续重叠的 n-grams。

```
>>> sent=['The', 'dog', 'gave', 'John', 'the', 'newspaper']
>>> n=3
>>> [sent[i:i+n] for i in range(len(sent)-n+1)]
[['The', 'dog', 'gave'], ['dog', 'gave', 'John'], ['gave', 'John', 'the'], ['John', 'the', 'newspaper']]
```

NLTK 提供了支持函数 bigrams(text)、trigrams(text)和一个更通用的 ngrams(text，n)。

```
>>> from nltk import trigrams
>>> sent=['The', 'dog', 'gave', 'John', 'the', 'newspape']
>>> trigrams(sent)
<generator object trigrams at 0x000000000780F048>
>>> list(trigrams(sent))
[('The', 'dog', 'gave'), ('dog', 'gave', 'John'), ('gave', 'John', 'the'), ('John', 'the', 'newspape')]
```

下面是使用循环变量构建多维结构的一个例子（from numpy import array）：建立一个 m 行 n 列的数组，其中每个元素是一个集合，可以使用一个嵌套的列表推导。

```
>>> import pprint
>>> r, c=3, 7
>>> array=[[set() for i in range(c)] for j in range(r)]
>>> array[2][5].add('Alice') #append, insert
>>> pprint.pprint(array)
[[set(), set(), set(), set(), set(), set(), set()],
 [set(), set(), set(), set(), set(), set(), set()],
 [set(), set(), set(), set(), set(), {'Alice'}, set()]]
>>> print(array)
[[set(), set(), set(), set(), set(), set(), set()], [set(), set(), set(), set(), set(), set(), set()], [set(), set(), set(), set(), set(), {'Alice'}, set()]]
```

循环变量 i 和 j 在产生对象过程中没有用到，它们只是需要一个语法正确的 for 语句。注意：由于有关对象复制的原因，使用乘法（复制）做这项工作是不正确的。

```
>>> import pprint
>>> r, c=3, 7 #3行7列, 37
>>> array=[[set()]*c]*r
>>> array[2][5].add(7)
>>> pprint.pprint(array)
[[{7}, {7}, {7}, {7}, {7}, {7}, {7}],
 [{7}, {7}, {7}, {7}, {7}, {7}, {7}],
 [{7}, {7}, {7}, {7}, {7}, {7}, {7}]]
>>> set()
set()
>>> ()
()
>>> {}
{}
>>> type({})
<class 'dict'>
>>> type(())
<class 'tuple'>
>>> type(set())
<class 'set'>
>>> [set()]*c
[set(), set(), set(), set(), set(), set(), set()]
>>> [[set()]*c]*r
[[set(), set(), set(), set(), set(), set(), set()], [set(), set(), set(), set(), set(), set()], [set(), set(), set(), set(), set(), set(), set()]]
```

4.3 函 数

函数提供了程序代码封装和重用的有效途径，假设我们发现经常要从 HTML 文件读取文本，包括打开文件、将它读入、规范化空白符号、剥离 HTML 标记等步骤，可以将这些步骤封装到一个函数中，并给它一个名字，如 get_text()，clean_html()，SB()。

从 HTML 文件中读取文本。

```
<html>
  <head>
    <title>
      测试
    </title>
  </head>
  <body>
    Hello, world!
  </body>
</html>
>>> from re import sub
>>> def get_text(file):
...     text=open(file).read()
...     text=sub('\s+', ' ', text)
...     text=sub('<.*?>', ' ', text)
...     return text
...
>>> get_text('file.html')
'    测试        Hello, world!    '
```

想从一个 HTML 文件得到纯文本，都可以用文件的名字作为唯一的参数调用 get_text()，它会返回一个字符串，我们可以将它赋值给一个变量 contents=get_text("file.html")。

每次要使用这一系列的步骤，只需要调用这个函数，使用函数可以为程序节约空间，更重要的是，为函数选择合适的名称可以提高程序的可读性。

函数不仅有助于提高重用性和可读性，还有助于提高程序的可靠性。

4.3.1 参数类型检查

写程序时 Python 不会强迫声明变量的类型，这允许我们定义参数类型灵活的函数。例如，下面 tag() 函数的作者假设其参数始终是一个字符串。

```
def tag(w):
    if w in ['a', 'the', 'all']:
        return 'det'
    else:
        return 'noun'

tag('the') #'det'
tag('knight') #'noun'
tag(["'Tis", 'but', 'a', 'scratch'])
>>> def tag(w):
...     if w in ['a', 'the', 'all']:
...         return 'det'
...     else:
...         return 'noun'
...
>>> tag('the') #'det'
'det'
>>> tag('knight') #'noun'
'noun'
>>> tag(["'Tis", 'but', 'a', 'scratch'])
'noun'
>>> w=["'Tis", 'but', 'a', 'scratch']
>>> w in ['a', 'the', 'all']
```

该函数对参数'the'和'knight'返回合理的值。传递给它一个列表，看看会发生什么？它没有报错，它返回的结果显然是不正确的。此函数的作者可以采取一些额外的步骤来确保 tag() 函数的参数 w 是一个字符串。一种做法是使用 if not type(w) is str 检查参数的类型，如果 w 不是一个字符串，简单地返回 Python 特殊的空值 None，这是一个较小的改善。该函数在检查参数类型，试图对错误的输入返回一个特殊的诊断结果，然而，它是有风险的，因为调用程序可能不会检测 None，所以这种诊断的返回值可能被传播到程序的其他部分，可能产生不可预测的后果。一个更好的解决方案是，使用 assert 语句和 Python 的 str 类型。

```
C:\Anaconda3>python
Python 3.4.1 |Anaconda 2.1.0 (64-bit)| (default, Sep 24 2014,
18:32:42) [MSC v.1600 64 bit (AMD64)] on win32
Type "help", "copyright", "credits" or "license" for more information.
>>> def tag(w):
...     assert isinstance(w, str), "argument to tag() must be a string"
...     if w in['a', 'the', 'all']:
...         return 'det'
```

```
...        else:
...            return 'noun'
...
>>> tag('the')
'det'
>>> tag('knight')
'noun'
>>> tag(["'Tis", 'but', 'a', 'scratch'])
Traceback (most recent call last):
  File "<stdin>", line 1, in <module>
  File "<stdin>", line 2, in tag
AssertionError: argument to tag() must be a string
>>> x=u'but'
>>> x
'but'
>>> type(x)
<class 'str'>
>>> repr(x)
"'but'"
```

如果 assert 语句失败，它会产生一个不可忽视的错误并停止程序执行。此外，该错误信息是容易理解的，程序中添加断言能帮助我们找到逻辑错误，是一种防御性编程。

4.3.2 函数的设计

结构良好的程序通常广泛使用函数。一个程序代码块超过 20 行时，如果将代码模块化成一个或多个函数，每一个函数有明确的功能，这将对可读性有很大的帮助。函数提供了一种重要的抽象，它让我们将多个操作组合成一个单一复杂的操作，并给它取一个名称，当我们使用函数时，主程序可以在一个更高的抽象层次编写，使其结构更清晰。适当地使用函数使程序更具可读性和可维护性。另外，重新实现一个函数已成为可能，使用更高效的代码替换函数体，不需要关心程序的其余部分。

freq_words 函数，它的作用是更新一个作为参数传递进来的频率分布的内容，并输出前 n 个最频繁词的列表。

```
C:\Anaconda3>python
Python 3.4.1 |Anaconda 2.1.0 (64-bit)| (default, Sep 24 2014,
18:32:42) [MSC v.1600 64 bit (AMD64)] on win32
Type "help", "copyright", "credits" or "license" for more information.
>>> from nltk import FreqDist, word_tokenize
```

```
>>> from bs4 import BeautifulSoup
>>> def freq_words(freqdist, n):
...     html=open('file.html').read()
...     soup=BeautifulSoup(html, "html")
...     text=soup.get_text()
...     for word in word_tokenize(text):
...         freqdist[word.lower()]+=1
...     print(freqdist.most_common()[:n])
...
>>> fd=FreqDist()
>>> freq_words(fd, 2)
[('.', 2), ('a', 2)]
>>> html=open('file.html').read()
>>> html
'<html>\n <head>\n  <title>\n  测试\n  </title>\n </head>\n <body>\nHello , world ! This a book . That is a desk .\n </body>\n</html>'
>>> soup=BeautifulSoup(html, "html")
>>> soup
<html>
<head>
<title>
    测试
   </title>
</head>
<body>
    Hello , world ! This a book . That is a desk .
  </body>
</html>
>>> repr(soup)
'<html>\n<head>\n<title>\n   测试 \n   </title>\n</head>\n<body>\nHello , world ! This a book . That is a desk .\n </body>\n</html>'
>>> text=soup.get_text()
>>> text
'\n\n\n  测试\n \n\n\n Hello , world ! This a book . That is a desk .\n \n'
>>> word_tokenize(text)
['测试', 'Hello', ',', 'world', '!', 'This', 'a', 'book', '.', 'That', 'is', 'a', 'desk', '.']
```

该函数有两个副作用：一是修改了第一个参数的内容；二是输出它已计算的结果是经过选择的子集。如果我们在函数内部初始化 FreqDist()对象，在它被处理的同一个地方，并且去掉选择集而将结果显示给调用程序，函数会更容易理解和在其他地方重用。

精心设计的函数用来计算高频词。

```
>>> from nltk import FreqDist, word_tokenize
>>> from bs4 import BeautifulSoup
>>> def freq_words():
...     freqdist=FreqDist()
...     html=open('file.html').read()
...     soup=BeautifulSoup(html, "html")
...     text=soup.get_text()
...     for word in word_tokenize(text):
...         freqdist[word.lower()]+=1
...     return freqdist
...
>>> fd=freq_words()
>>> print(fd.most_common()[:2])
[('.', 2), ('a', 2)]
```

4.3.3 作为参数的函数

Python 允许我们传递一个函数作为另一个函数的参数，现在，我们可以抽象出操作，对相同数据进行不同操作。例如，可以传递内置函数 len()或用户定义的函数 last_letter()作为另一个函数的参数。

```
>>> sent='Take care of the sense , and the sounds will take care of themselves .'.split()
>>> def extract_property(prop): #单词的特征提取
...     return [prop(word) for word in sent]
...
>>> extract_property(len) #单词的特征提取，长度
[4, 4, 2, 3, 5, 1, 3, 3, 6, 4, 4, 4, 2, 10, 1]
>>> def last_letter(word):
...     return word[-1]
...
>>> extract_property(last_letter) #单词的特征提取，最后一个字符
['e', 'e', 'f', 'e', 'e', ',', 'd', 'e', 's', 'l', 'e', 'e', 'f', 's', '.']
```

函数len()和last_letter()可以像列表和字典那样被传递。注意，只有在调用该函数时，才在函数名后使用括号。如果只是将函数作为一个参数，括号将被省略。Python提供了更多的方式来定义函数作为其他函数的参数，即lambda表达式。

传递一个函数给sorted()函数。可以提供自己的排序函数，例如按长度递减排序。

```
>>> def extract_property(prop):
...     return [prop(word) for word in sent]
...
>>> extract_property(len)
[4, 4, 2, 3, 5, 1, 3, 3, 6, 4, 4, 4, 2, 10, 1]
>>> def last_letter(word):
...     return word[-1]
...
>>> extract_property(last_letter)
['e', 'e', 'f', 'e', 'e', ',', ',', 'd', 'e', 's', 'l', 'e', 'e', 'f', 's', '.']
>>> extract_property(lambda w:w[-1])
['e', 'e', 'f', 'e', 'e', ',', ',', 'd', 'e', 's', 'l', 'e', 'e', 'f', 's', '.']
>>> sorted(sent)
[',', '.', 'Take', 'and', 'care', 'care', 'of', 'of', 'sense', 'sounds', 'take', 'the', 'the', 'themselves', 'will']
>>> sorted(sent, key=lambda x:len(x))
[',', '.', 'of', 'of', 'the', 'and', 'the', 'Take', 'care', 'will', 'take', 'care', 'sense', 'sounds', 'themselves']
```

4.3.4 迭代与递归

这些函数从初始化一些存储开始，迭代和处理输入的数据，最后返回最终的对象，一个大的结构或汇总的结果。一个标准的方式是初始化一个空列表，累计材料，然后返回这个列表。

累计输出词列表words中含substring的单词到一个列表。

```
>>> def search1(substring, words):
...     result=[]
...     for word in words:
...         if substring in word:
...             result.append(word)
...     return result  #一次性全部求出
...
```

```
>>> def search2(substring, words):
...     for word in words:
...         if substring in word:
...             yield word #按需求出
...
>>> words='That is a desk .'
>>> words=words.split()
>>> print "search1:"
search1:
>>> #累计输出'That is a desk .'中含'a'的单词到一个列表
... #['That', 'a']
... for item in search1('a', words):
...     print item
...
That
a
>>> print "search2:"
search2:
>>> for item in search2('a', words):
...     print item
...
That
a
```

函数 search2() 是一个产生器，第一次调用此函数，它运行到 yield 语句然后停下来，调用程序获得第一个词，一旦调用程序对另一个词做好准备，函数会从停下来的地方继续执行，直到再次遇到 yield 语句。这种方法通常更有效，因为函数只产生调用程序需要的数据，并不需要分配额外的内存，如 result=[]存储输出（正如语法分析程序与词法分析程序一样）。

下面是一个更复杂的产生器的例子，产生一个词列表的所有排列，为了强制 permutations() 函数产生所有的输出，将它包装在 list() 调用中。

```
>>> def permutations(seq): #产生1, 2, 3所有排列
...     if len(seq)<=1:
...         yield seq
...     else: #递归产生2, 3所有排列: 2 3 ; 3 2
...         for perm in permutations(seq[1:]):
...             for i in range(len(perm)+1):
...                 #对2, 3每一个排列插入1的位置i=0, 1, 2
```

```
...                      yield perm[:i]+seq[0:1]+perm[i:]# 列表加,
seq[0:1]seq[0]
...
>>> list(permutations(['police', 'fish', 'buffalo']))
[['police', 'fish', 'buffalo'], ['fish', 'police', 'buffalo'],
['fish', 'buffalo', 'police'], ['police', 'buffalo', 'fish'],
['buffalo', 'police', 'fish'], ['buffalo', 'fish', 'police']]
>>> list(permutations([1, 2, 3]))
[[1, 2, 3], [2, 1, 3], [2, 3, 1], [1, 3, 2], [3, 1, 2], [3, 2, 1]]
```

permutations 函数使用了递归技术，产生一组词的排列对于创建测试一个语法的数据十分有用。

4.3.5　filter 和 map 函数

Python 提供了一些具有函数式编程语言标准特征的高阶函数，我们将在这里演示它们。从定义一个函数 is_content_word()开始，它检查一个词是否来自一个开放的实词类，使用此函数作为 filter()的第一个参数，它对作为它的第二个参数的序列中的每个项目运用该函数，只保留该函数返回 True 的项目。

```
>>> def is_content_word(word):
...     return word.lower() not in ['a', 'of', 'the', 'and', 'will',
',', '.']
...
>>> sent=['Take', 'care', 'of', 'the', 'sense', ',', 'and', 'the',
'sounds', 'will', 'take', 'care', 'of', 'themselves', '.'] #注意你的
理智,声调自会小心。
>>> filter(is_content_word, sent)
['Take', 'care', 'sense', 'sounds', 'take', 'care', 'themselves']
>>> [w for w in sent if is_content_word(w)]
['Take', 'care', 'sense', 'sounds', 'take', 'care', 'themselves']
>>> word='Take'
>>> word.lower()
'take'
>>> word.lower() not in ['a', 'of', 'the', 'and', 'will', ',', '.']
True
>>> word='The'
>>> word.lower() not in ['a', 'of', 'the', 'and', 'will', ',', '.']
False
```

```
>>> [is_content_word(w) for w in sent]
[True, True, False, False, True, False, False, False, True, False,
True, True, False, True, False]
>>> ' '.join(['Take', 'care', 'of', 'the', 'sense', ',', 'and',
'the', 'sounds', 'will', 'take', 'care', 'of', 'themselves', '.'])
'Take care of the sense , and the sounds will take care of themselves .'
```

高阶函数 map()的作用是，将一个函数运用到一个序列中的每一项。这里是一个找出 brown 语料库 news 部分中句子的平均长度的方法，后面跟着的是使用列表推导计算的等效版本。

```
>>> from nltk.corpus import brown
>>> lengths=list(map(len, brown.sents(categories='news')))
>>> sum(lengths)/len(lengths) #21.7508111616
21.75081116158339
>>> lengths=[len(w) for w in brown.sents(categories='news')]
>>> sum(lengths)/len(lengths) #21.7508111616
21.75081116158339
```

在上面的例子中，我们指定了一个用户定义的函数 is_content_word()和一个内置函数 len()，这里还可以提供一个 lambda 表达式，计数每个词中的元音的数量（这是两个等效的例子）。

```
>>> sent=['Take', 'care', 'of', 'the', 'sense', ',', 'and', 'the',
'sounds', 'will', 'take', 'care', 'of', 'themselves', '.']
>>> list(map(lambda w:len(list(filter(lambda c:c.lower() in
"aeiou", w))), sent))
[2, 2, 1, 1, 2, 0, 1, 1, 2, 1, 2, 2, 1, 3, 0]
>>> [len([c for c in w if c.lower() in "aeiou"]) for w in sent]
[2, 2, 1, 1, 2, 0, 1, 1, 2, 1, 2, 2, 1, 3, 0]
>>> [2, 2, 1, 1, 2, 0, 1, 1, 2, 1, 2, 2, 1, 3, 0]
[2, 2, 1, 1, 2, 0, 1, 1, 2, 1, 2, 2, 1, 3, 0]
>>> [[c for c in w if c.lower() in "aeiou"] for w in sent]
[['a', 'e'], ['a', 'e'], ['o'], ['e'], ['e', 'e'], [], ['a'], ['e'],
['o', 'u'], ['i'], ['a', 'e'], ['a', 'e'], ['o'], ['e', 'e', 'e'], []]
```

列表推导是基础的解决方案，它通常比基于高阶函数的解决方案的可读性高。

4.3.6 参数的命名

当有很多参数时，很容易混淆正确的顺序，这时可以通过名字引用参数，甚至可以给它们分配默认值以供调用程序没有提供该参数时使用。现在，参数可以按任意顺序指定，也可以省略。

```
>>> def repeat(msg='<empty>', num=1):
...     return msg*num
...
>>> repeat(num=3)
'<empty><empty><empty>'
>>> repeat(msg='Alice')
'Alice'
>>> repeat(num=5, msg='Alice')
'AliceAliceAliceAliceAlice'
```

这些称为关键字参数，如果混合使用这两种参数，就必须确保未命名的参数在命名的参数前面，因为未命名参数是根据位置来定义的。

可以定义一个函数，接受任意数量的未命名和命名参数，并通过一个参数列表*args 和一个参数字典**kwargs 来访问它们。

```
>>> def generic(*args, **kwargs):
...     print args
...     print kwargs
...
>>> generic(1, "African swallow", monty="python")
(1, 'African swallow')
{'monty': 'python'}
>>> def generic(*args, **kwargs):
...     print args
...     print kwargs
...     print type(args)
...     print type(kwargs)
...
>>> generic(1, "African swallow", monty="python")
(1, 'African swallow')
{'monty': 'python'}
<type 'tuple'>
<type 'dict'>
```

4.4 程序开发

4.4.1 Python 模块的结构

程序模块的目的是把逻辑上相关的定义和函数结合在一起，以方便重用和更高层次的抽象。Python 模块是一些单独的.py 文件，例如，在处理一种特定的语料格式时，读取和写入这种格式的函数可以放在一起，这种格式所使用的常量，如字段分隔符或一个 EXTN=".inf" 文件扩展名，可以共享。如果要更新格式，只有一个文件需要改变。同样地，一个模块可以包含用于创建和操纵一种特定的数据结构（如语法树的代码），或执行特定的处理任务（如绘制语料统计图表的代码）。开始编写 Python 模块时，可以使用变量__file__定位系统中任一 NLTK 模块的代码。

```
>>> from nltk.metrics import distance
>>> distance.__file__
'C:\\Users\\LYS\\Anaconda3\\lib\\site-packages\\nltk\\metrics\\distance.py'
```

将返回模块文件的位置，不同的设备可能看到的位置不同。需要打开的文件是对应的.py 源文件，它和.pyc 文件放在同一目录下。与其他 NLTK 的模块一样，distance.py 以一组注释行开始，包括一行模块标题和作者信息，代码会被发布，包括代码可用的 URL、版权声明和许可信息。接下来是模块级的 docstring，三重引号的多行字符串，其中包括输入 help(distance) 时将被输出的模块信息。

```
>>> help(distance)
Help on module nltk.metrics.distance in nltk.metrics:

NAME
    nltk.metrics.distance - Distance Metrics.

DESCRIPTION
    Compute the distance between two items (usually strings).
    As metrics, they must satisfy the following three requirements:

    1. d(a, a) = 0
    2. d(a, b) >= 0
    3. d(a, c) <= d(a, b) + d(b, c)

FUNCTIONS
    binary_distance(label1, label2)
        Simple equality test.

        0.0 if the labels are identical, 1.0 if they are different.

        >>> from nltk.metrics import binary_distance
        >>> binary_distance(1, 1)
        0.0

        >>> binary_distance(1, 3)
-- More --
```

在这之后是模块的导入语句，然后是全局变量，接着是组成模块主要部分的一系列函数的定义。有些模块定义"类"（面向对象编程的主要部分），大多数 NLTK 的模块还包括一个 demo()函数，可以用来演示模块使用方法。模块的一些变量和函数仅用于模块内部，它们的名字应该以下划线开头，如_helper()。如果使用 from module import *导入这个模块，这些名称将不会被导入，可以选择性地列出一个模块的外部可访问的名称，此时使用一个特殊的内置变量__all__=['edit_distance', 'jaccard_distance']。

4.4.2　Pdb 调试器

Python 提供了一个调试器，用于监视程序的执行，指定程序暂停运行的行号（即断点），逐步调试代码段和检查变量的值。可以使用如下方式在代码中调用调试器。

```
import pdb
import mymodule
pdb.run('mymodule.myfunction()')
```

它会给出一个提示(Pdb)，此时可以在那里输入指令给调试器。输入 help 可查看命令的完整列表；输入 step 或只输入 s 将执行当前行然后停止。如果当前行调用一个函数，它将进入这个函数并停止在第一行；输入 next 或只输入 n，它会在当前函数中的下一行停止执行；break 或 b 命令可用于创建或列出断点；输入 continue 或 c 会继续执行直到遇到下一个断点，输入任何变量的名称可以检查它的值。

可以使用 Python 调试器来查找 find_words()函数的问题。问题一般是在第二次调用函数时产生的，一开始可以不使用调试器而调用该函数。使用可能的最小输入，在第二次使用调试器时调用它。

```
def find_words(text, wordlength, result=[]):
    for word in text:
        if len(word)==wordlength:
            result.append(word)
    return result

import pdb
find_words(['cat'], 3)

pdb.run("find_words(['dog'], 3)")
><string>(1)<module>()
(Pdb)step
--Call--
><stdin>(1)find_words()
(Pdb)args
```

```
text=['dog']
wordlength=3
result=['cat']
```

在这里,只输入了两个命令到调试器:step 带入函数体内部,args 显示它的参数值。此时可看到 result 有一个初始值['cat'],而不是如预期的空列表,调试器帮助定位问题,促使检查我们对 Python 函数的理解。

4.5 算法设计

4.5.1 递归与迭代

解决一个大小为 n 的问题,可以将其分成两半,然后处理一个或两个大小为 $n/2$ 的问题,一般的方式是使用迭代。定义一个函数 f,简化问题,调用自身来解决一个或两个同样问题更简单的实例,然后组合它们的结果成为原问题的解答。假设有 n 个词,要计算它们结合在一起有多少不同的方式能组成一个词序列。如果只有 1 个词,只有 1 种方式组成一个序列,如果有 2 个词,就有 2 种方式将它们组成一个序列,3 个词有 6 种可能,一般地,n 个词有 $n \times (n-1) \times \cdots \times 2 \times 1$ 种方式,即 n 的阶乘 $n!$。可以将这些编写成如下代码:

```
>>> def factorial1(n):
...     result=1
...     for i in range(n):  #迭代算法
...         result*=i+1    #累乘器
...     return result
...
>>> factorial1(1)
1
>>> factorial1(2)
2
>>> factorial1(3)
6
```

也可以使用另外一种算法来解决这个问题。该算法基于以下观察,假设有办法为 $n-1$ 个不同的词构建所有的排列,然后对于每个这样的排列,有 n 个地方可以插入一个新词,开始、结束或任意两个词之间的 $(n-1)-1$ 个空隙,因此,简单地将 $n-1$ 个词的解决方案数乘以 $2+(n-1)-1=n$ 的值,$n!=n \times (n-1)!$。我们还需要基础案例,也就是说,如果有一个词只有一个顺序,可以将这些编写成如下代码:

```
>>> def factorial2(n):
...     if n==1:
```

```
...         return 1
...     else:
...         return n*factorial2(n-1)  #递归算法
...
>>> factorial2(1)
1
>>> factorial2(2)
2
>>> factorial2(3)
6
```

这两种算法能解决同样的问题,一个使用迭代,而另一个使用递归。可以用递归处理深层嵌套的对象,例如 WordNet 的上位词层次。计数给定同义词集 s 为根的上位词层次的大小(即以 s 为根的树的结点数),会找到 s 的每个下位词的大小,然后将它们加到一起,将加 1 表示同义词集本身。下面的函数 size1()可以做这项工作,注意函数体中包括 size1()的递归调用。

```
>>> from nltk.corpus import wordnet as wn
>>> def size1(s):
...     return 1+sum(size1(child) for child in s.hyponyms())
...
>>> dog=wn.synset('dog.n.01')
>>> size1(dog)
190
```

可以设计一种这个问题的迭代解决方案处理层的层次结构。第一层是同义词集本身,然后是同义词集所有的下位词,之后是所有下位词的下位词,每次循环通过查找上一层的所有下位词计算下一层,它也保存了到目前为止遇到的同义词集的总数(同义词集即结点,但不是严格的树)。

```
>>> from nltk.corpus import wordnet as wn
>>> def size2(s):
...     layer=[s]  #队列,层序遍历
...     total=0
...     while layer:
...         total+=len(layer)
...         layer=[h for c in layer for h in c.hyponyms()]
...     return total
...
>>> dog=wn.synset('dog.n.01')
>>> size2(dog)
190
```

迭代解决方案不仅代码更长而且更难理解，它迫使我们程序式地思考问题并跟踪 layer 和 total 随时间变化发生了什么。这两种解决方案均给出了相同的结果。

下面用递归算法构建一个深嵌套的对象。一个字母查找树是一种可以用来索引词汇的数据结构，一次一个字母，根据词检索 word retrieval 而得名。例如，如果 trie 包含一个字母的查找树，那么 trie['c']是一个较小的查找树，包含所有以 c 开头的词。下面演示使用 Python 字典构建查找树的递归过程。插入词 chien，将 c 分类，递归地掺入 hien 到 trie['c']的子查找树中，递归处理直到词中没有剩余的字母，存储预期值。

构建一个字母查找树。一个递归函数建立一个嵌套的字典结构，每一级嵌套包含给定前缀的所有单词，子查找树含有所有可能的后续词。

```
>>> from pprint import pprint
>>> def insert(trie, key, value):
...     if key:
...         first, rest=key[0], key[1:]
...         if first not in trie:
...             trie[first]={}
...         insert(trie[first], rest, value)
...     else:
...         trie['value']=value
...
>>> trie={} #nltk.defaultdict(dict)
>>> insert(trie, 'chat', 'cat') #insert(trie, key, value)
>>> insert(trie, 'chien', 'dog')
>>> insert(trie, 'chair', 'flesh')
>>> insert(trie, 'chic', 'stylish')
>>> trie=dict(trie) #for nicer printing
>>> trie['c']['h']['a']['t']['value'] #'cat'
'cat'
>>> pprint(trie)
{'c': {'h': {'a': {'i': {'r': {'value': 'flesh'}}, 't': {'value': 'cat'}},
            'i': {'c': {'value': 'stylish'}, 'e': {'n': {'value': 'dog'}}}}}}
```

尽管递归编程结构简单，但它是有代价的。每次调用函数时，一些状态信息需要推入堆栈，这样一旦函数执行完成可以从离开的地方继续执行。出于这个原因，迭代解决方案往往比递归解决方案更高效。

4.5.2 权衡空间与时间

通过建设一个辅助的数据结构（如索引），有时可以显著地加快程序的执行。下面实现了一个简单电影评论语料库的全文检索系统，通过索引文档集合，它可以提供较快的查找。

一个简单的全文检索系统。

```
C:\Anaconda3>python
Python 3.4.1 |Anaconda 2.1.0 (64-bit)| (default, Sep 24 2014,
18:32:42) [MSC v.1600 64 bit (AMD64)] on win32
Type "help", "copyright", "credits" or "license" for more information.
>>> from nltk import Index
>>> from re import sub
>>> from nltk.corpus import movie_reviews
>>> def raw(file):
...     contents=open(file).read()
...     contents=sub('<.*?>', ' ', contents) #clean_html
...     contents=sub('\s+', ' ', contents)
...     return contents
...
>>> def snippet(doc, term): #buggy
...     text=' '*30+raw(doc)+' '*30
...     pos=text.index(term)
...     return text[pos-30:pos+30]
...
>>> print("Building Index ...")
Building Index ...
>>> files=movie_reviews.abspaths()
>>> idx=Index((w, f) for f in files for w in raw(f).split())
>>> #全文索引,,倒排表,{'This': ['file1.txt', 'file2.txt'], 'is': ['file1.txt']}
... query=''
>>> while query!="quit":
...     query=input("query>")
...     if query in idx:
...         for doc in idx[query]:
...             print(snippet(doc, query))
...     else:
...         print("Not found")
...
```

```
query>nice
, cool music , claire dane's nice hair and cute outfits , c
erican roles " for you as a " nice guy " in a romantic comed
dwin , but it might have been nice to have learned something
...
```

sub 是替换，<是普通字符，.是匹配任意字符（除换行符外），*是匹配前面的任意字符，?是非贪婪，>是普通字符，组合起来是将<、中间的任意字符和>换为空字符串。因为有？（是非贪婪），所以是匹配<后面最近的一个>为止，功能是去除 HTML 标记。

```
>>> x=[('This', 'file1.txt'), ('is', 'file1.txt'), ('This', 'file2.txt')]
>>> Index(x)
defaultdict(<class 'list'>, {'This': ['file1.txt', 'file2.txt'],
'is': ['file1.txt']})
```

倒排表，表示 file1.txt 和 file2.txt 文件中都有 This。根据 This，可以索引文件 file1.txt 和 file2.tx，即全文索引。

```
>>> x='This is a book'
>>> pos=x.index('is') #2
>>> x[pos-1:pos+1] #hi
'hi'
```

一个更微妙的空间与时间折中的例子涉及使用整数标识符替换一个语料库的标识符。为语料库创建一个词汇表，每个词都被存储一次的列表，然后转化这个列表以便能通过查找任意词来找到它的标识符。每个文档都进行预处理，使一个词列表变成一个整数列表，现在所有的语言模型都可以使用整数。

预处理已标注的语料库数据，将所有的词和标注转换成整数。

```
>>> from nltk.corpus import brown
>>> def preprocess(tagged_corpus):
...     words=set()
...     tags=set()
...     for sent in tagged_corpus:
...         for word, tag in sent:
...             words.add(word)
...             tags.add(tag)
...     wm=dict((w, i) for (i, w) in enumerate(words))
...     tm=dict((t, i) for (i, t) in enumerate(tags))
...     return [[(wm[w], tm[t]) for (w, t) in sent] for sent in
```

```
tagged_corpus]
...
>>> tagged_corpus =brown.tagged_sents()[:1]
>>> preprocess(tagged_corpus)
[[(10, 14), (11, 9), (24, 10), (17, 8), (8, 10), (16, 3), (9, 7),
(2, 14), (6, 0), (7, 1), (1, 5), (20, 11), (13, 0), (21, 0), (12, 3),
(18, 12), (0, 14), (19, 0), (22, 4), (15, 15), (3, 2), (5, 13), (4, 3),
(14, 0), (23, 6)]]
```

空间与时间权衡的另一个例子是维护一个词汇表。如果需要处理一段输入文本，检查所有的词是否在现有的词汇表中，词汇表应存储为一个集合，而不是一个列表，集合中的元素会自动索引，所以测试一个大集合的成员将远远快于测试相应列表的成员。可以使用 timeit 模块进行测试。Timer 类有两个参数，一个是多次执行的语句，一个是只在开始执行一次的设置代码，分别使用一个 list 和一个 set 模拟 10 万个项目的词汇表。测试语句将产生一个随机项，它有 50%的机会在词汇表中。

```
>>> from timeit import Timer
>>> setup_list = "import random; vocab = list(range(100000))" #
强制类型转换
>>> setup_set = "import random; vocab = set(range(100000))"
>>> statement = "random.randint(0, 200000) in vocab"
>>> Timer(statement, setup_list).timeit(1000)
0.7351529000000028
>>> Timer(statement, setup_set).timeit(1000)
0.0012291000000033137
vocab_size=100000
import random
vocab=range(vocab_size)
for i in range(1000):
    _=random.randint(0, vocab_size*2) in vocab

vocab_size = 100000
import random
vocab = set(range(vocab_size))
for i in range(1000):
    _=random.randint(0, vocab_size*2) in vocab
```

执行 1 000 次列表成员资格测试总共需要 0.735 152 900 000 002 8 s≈1 s，而在集合上的等效试验仅需 0.001 229 100 000 003 313 7 s≈0.001 s，也就是说快了 3 个数量级。

4.6　Python 库的样例

4.6.1　brown 语料库中特殊情态动词的频率

Python 有一些可视化语言数据的库，Matplotlib 包支持 MATLAB 接口的绘图函数，以图形的形式显示数值数据往往是非常有用的，因为这往往更容易检测到模式。例如，一个图形显示的彩色图，显示按类别划分 brown 语料库中的特殊情态动词的频率，输出显示如图 4-1 所示。

brown 语料库中不同部分的情态动词频率。

```
>>> from nltk import ConditionalFreqDist
>>> from nltk.corpus import brown
>>> colors='rgbcmyk' #red, green, blue, cyan, magenta, yellow, black
>>> def bar_chart(categories, words, counts):
...     from pylab import arange, bar, xticks, legend, ylabel, title, show, rcParams
...     rcParams['font.sans-serif']=['SimHei'] #设置图形里的中文为黑体
...     ind=arange(len(words))
...     width=1.0/(len(categories)+1) #这里一定要注意用 1.0
...     bar_groups=[]
...     for c in range(len(categories)):
...         bars=bar(ind+c*width, counts[categories[c]], width, color=colors[c%len(colors)])
...         bar_groups.append(bars)
...     xticks(ind+width, words)
...     legend([b[0] for b in bar_groups], categories, loc='upper left')
...     ylabel(u'频率')
...     title(u'brown 语料库中不同部分的情态动词频率')
...     show()
...
>>> genres=['news', 'religion', 'hobbies', 'government', 'adventure']
>>> modals=['can', 'could', 'may', 'might', 'must', 'will']
>>> cfdist=ConditionalFreqDist((genre, word) for genre in genres for word in brown.words(categories=genre) if word in modals) #分组统计
>>> counts={}
>>> for genre in genres:
...     counts[genre]=[cfdist[genre][word] for word in modals]
...
>>> bar_chart(genres, modals, counts)
```

第 4 章 结构化程序设计

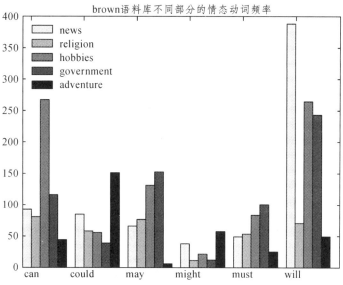

图 4-1　brown 语料库中不同部分的情态动词频率

从该柱状图可以看出，may 和 must 有几乎相同的相对频率，could 和 might 也一样。

可以动态地产生这些数据的可视化图形。要做到这一点，必须为 matplotlib 指定 Agg 后台，它是一个产生栅格（像素）图像的库。下一步，像以前一样使用相同的 PyLab 方法，但不是用 pylab.show() 显示结果在图形终端，而是使用 pylab.savefig() 把它保存到一个文件。注意，matplotlib.use('agg') 必须在本句执行 import matplotlib.pyplot as plt 前运行。

```
from numpy import arange, sin, pi
from matplotlib.pyplot import clf, plot, axhline, axvline, show
x = arange(0, 2*pi, 0.001)
y = sin(2 * pi * x)
clf()
plot(x, y)
l = axhline(linewidth=1, color='black')
l = axvline(linewidth=1, color='black')
show()
from numpy import arange, sin, pi
import matplotlib
from pylab import savefig
matplotlib.use('agg')
from matplotlib.pyplot import clf, plot, axhline, axvline, show
x = arange(0, 2*pi, 0.001)
y = sin(2 * pi * x)
clf()
```

```
plot(x, y)
l = axhline(linewidth=1, color='black')
l = axvline(linewidth=1, color='black')
savefig('modals.png')
```

4.6.2　WordNet 网络结构的可视化

Anaconda-2.1.0-Windows-x86_64 集成了 Network X 1.9.1。Network X 可以和 Matplotlib 结合使用可视化 WordNet 等的网络结构。下面示例中的程序初始化一个空的图，然后遍历 WordNet 上位词层次为图添加边。注意，遍历是递归的。结果显示如图 4-2 所示。

使用 Network X 和 Matplotlib 库。

```
Python 2.7.8 |Anaconda 2.1.0 (64-bit)| (default, Jul  2 2014, 
15:12:11) [MSC v.1500 64 bit (AMD64)] on win32
Type "help", "copyright", "credits" or "license" for more information.
Anaconda is brought to you by Continuum Analytics.
Please check out: http://continuum.io/thanks and https://binstar.org
>>> import networkx as nx
>>> import matplotlib
>>> from nltk.corpus import wordnet as wn
>>> def traverse(graph, start, node):
...     graph.depth[node.name]=node.shortest_path_distance(start)
...     for child in node.hyponyms():
...         graph.add_edge(node.name, child.name)
...         traverse(graph, start, child)
...
>>> def hyponym_graph(start):
...     G=nx.Graph()
...     G.depth={}
...     traverse(G, start, start)
...     return G
...
>>> def graph_draw(graph):
...     nx.draw_graphviz(graph, node_size=[16*graph.degree(n) for n in graph], node_color=[graph.depth[n] for n in graph], with_labels=False)
...     #Network X 1.9.1有 draw_graphviz, 高版本没有 draw_graphviz
...     matplotlib.pyplot.show()
...
>>> dog=wn.synset('dog.n.01')
>>> graph=hyponym_graph(dog)
>>> graph_draw(graph)
```

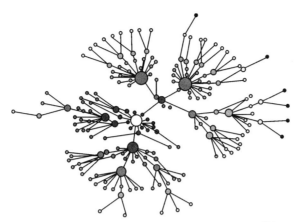

图 4-2　使用 Network X 和 Matplotlib 可视化数据

WordNet 上位词层次的部分显示开始于 dog.n.01（中间最黑的节点），节点的大小对应节点的孩子的数目，颜色对应节点到 dog.n.01 的距离。

4.6.3　CSV 格式的简单词典

语言分析工作往往涉及数据统计表，包括有关词项的信息、试验研究的参与者名单或从语料库提取的语言特征。这里有一个简单的 CSV（Comma-Separated Values，逗号分隔型取值格式）词典片段。

```
sleep, sli:p, v.i, a condition of body and mind ...
walk, wo:k, v.intr, progress by lifting and setting down each foot ...
wake, weik, intrans, cease to sleep
```

可以使用 Python 的 CSV 库读写这种格式存储的文件。例如，可以打开一个叫作 lexicon.csv 的 CSV 文件，并遍历它的行。

```
>>> import csv
>>> input_file=open("lexicon.csv", "rb")
>>> for row in csv.reader(input_file):
...     print row
...
['sleep', ' sli:p', ' v.i', ' a condition of body and mind ...']
['walk', ' wo:k', ' v.intr', ' progress by lifting and setting down each foot ...']
['wake', ' weik', ' intrans', ' cease to sleep']
```

每一行是一个字符串列表，如果字段包含有数值数据，它们将作为字符串出现，所以必须使用 int() 或 float() 进行转换。

小 结

（1）Python 赋值和参数传递使用对象引用，例如，如果 a 是一个列表，赋值 b=a，然后 a 上的任何操作都将修改 b，反之亦然。

（2）is 操作测试是否两个对象是相同的内部对象，而==测试是否两个对象是相等的。两者的区别和标识符与类型的区别相似。

（3）字符串、列表和元组是不同类型的序列对象，支持常见的操作，如索引、切片、len()、sorted()和使用 in 的成员测试。

（4）可以通过写方式打开文件来写入文本到一个文件：ofile=open('output.txt', 'w')，然后加入内容到文件：ofile.write("Hello, world!")，最后关闭文件 ofile.close()。

（5）声明式的编程风格通常会产生更简洁、可读的代码，手动递增循环变量通常是不必要的，枚举一个序列，使用 enumerate()。

（6）函数是一个重要的编程抽象，需要理解的关键概念有参数传递、变量的范围和 docstrings。

（7）函数作为一个命名空间，函数内部定义的名称在该函数外不可见，除非这些名称被定义是全局的。

（8）模块允许将材料与本地的文件逻辑关联起来，一个模块作为一个命名空间，在一个模块中定义的名称，如变量和函数，在其他模块中不可见，除非这些名称被导入。

（9）动态规划是一种在 NLP 中广泛使用的算法设计技术，它存储以前的计算结果，以避免不必要的重复计算。

练 习

1. 使用 Python 的帮助功能，找出更多关于序列对象的内容。在解释器中输入 help(str)，help(list)和 help(tuple)。系统会给出一个每种类型支持的函数的完整列表。一些函数名字有些特殊，两侧有下划线，例如，x.__getitem__(y)仅仅是以长篇大论的方式使用 x[y]。

```
>>> help(str)
```

```
>>> dir(str)
```

2. 确定三个同时在元组和列表上都可以执行的操作，确定三个不能在元组上执行的列表操作，命名一段使用列表替代元组会产生一个 Python 错误的代码。

索引、排序(sorted())、切片和长度：在元组和列表上都可以执行的操作，读操作，不改变结构的操作。

修改、排序(sort())、删除：不能在元组上执行的列表操作，写操作，改变结构的操作。

```
lexicon=['the', 'off']
lexicon.sort()
```

```
lexicon
lexicon[1]='turned'
lexicon
del lexicon[0]
lexicon

lexicon=('turned', 'VBD')
lexicon.sort()
lexicon
lexicon[0]='the'
lexicon
del lexicon[0]
lexicon
```

3. 了解如何创建一个由单个项目组成的元组（至少有两种方法可以做到）。

```
TO BE CONTINUE…
x='snark',
y=('snark', )
z=tuple(['snark'])
```

4. 创建一个列表 words=['is', 'NLP', 'fun', '?']。使用一系列赋值语句（如 words[1]=words[2]）和临时变量 tmp 将这个列表转换为列表['NLP', 'is', 'fun', '!']。现在，使用元组赋值做相同的转换。

```
words=['is', 'NLP', 'fun', '?']
tmp=words[0]
words[0]=words[1]
words[1]=tmp
words[3]='!'
['NLP', 'is', 'fun', '!']
words=['is', 'NLP', 'fun', '?']
words[0], words[1]=words[1], words[0]   #构成一个圈
words[3]='!'
['NLP', 'is', 'fun', '!']
words=['is', 'NLP', 'fun', '?']
words=tuple(words)
tmp=words[0]
words[0]=words[1]
words[1]=tmp
words[3]='!'
```

```
Traceback (most recent call last):
  File "<stdin>", line 1, in <module>
TypeError: 'tuple' object does not support item assignment
>>> words[1]=tmp
Traceback (most recent call last):
  File "<stdin>", line 1, in <module>
TypeError: 'tuple' object does not support item assignment
>>> words[3]='!'
Traceback (most recent call last):
  File "<stdin>", line 1, in <module>
TypeError: 'tuple' object does not support item assignment
```

5. 通过输入 help(cmp)阅读关于内置的比较函数 cmp 的内容。它与比较运算符在行为上有何不同？

```
>>> help(cmp)
Help on built-in function cmp in module __builtin__:

cmp(...)
    cmp(x, y) → integer

    Return negative if x<y, zero if x==y, positive if x>y.


>>>
cmp(1, 2)
1<2
```

6. 创建一个 n-grams 滑动窗口的方法，在下面两种极端情况下是否正确：n=1 和 n=len(sent)？

```
sent=['The', 'dog', 'gave', 'john', 'the', 'newspaper']
n=1
[sent[i:i+n] for i in range(len(sent)-n+1)]
n=len(sent)
[sent[i:i+n] for i in range(len(sent)-n+1)]
正确
[['the'], ['dog'], ['gave'], ['john'], ['the'], ['newspaper']]
[['the', 'dog', 'gave', 'john', 'the', 'newspaper']]
```

7. 指出当空字符串和空列表出现在 if 从句的条件部分时，它们的判断结果是 False。在这种情况下，它们被说成出现在一个布尔上下文中。实验各种不同的布尔上下文中的非布尔表达式，看它们是否被判断为 True 或 False。

```
mixed=['cat', '', ['dog'], []]
for element in mixed:
    if element:
        print element
cat
['dog']
```

8. 使用不等号比较字符串，如'Monty'<'Python'。当你做'Z'<'a'时会发生什么？尝试具有共同前缀的字符串对，如'Monty'<'Montague'。阅读有关"字典排序"的内容以便了解这里发生了什么事。尝试比较结构化对象，如('Monty', 1)<('Monty', 2)。这与预期一样吗？

```
'Monty'<'Python'  #True
'Z'<'a'  #True
('Monty', 1)<('Monty', 2)  #True
```

9. 写代码删除字符串开头和结尾处的空白，并规范化词之间的空格为一个单独的空格字符。
（1）使用 split()和 join()。

```
s='\tNormalizing whitespace\t\n\r\f\v'
' '.join(s.split())
```

（2）使用正则表达式替换。

```
import re
s='\tNormalizing whitespace\t\n\r\f\v'
re.sub('^\s+|\s+$', '', s)
chr(ord('\n'))

import re
string='  Where       are we going to visit today    '
re.findall('\S+', string)
import re
string='  Where       are we going to visit today    '
string=re.sub('^ +| +$', '', string)
string=re.sub(' +', ' ', string)
9999999999999999
string=' Where       are we going to visit today    '
string=string.strip()
string=string.replace('  ', ' ')
```

10. 写一个程序按长度对词排序。定义一个辅助函数 cmp_len，它在词长上使用 cmp 比较函数。

```
def cmp_len(x, y):
    return cmp(len(x), len(y))

words='I turned off the spectroroute'.split()
words.sort(cmp=cmp_len)
words='I turned off the spectroroute'.split()
wordlens=[(len(word), word) for word in words]
wordlens.sort()
' '.join(w for _, w in wordlens)
words=['I', 'love', 'NPL']
words.sort(key=len)
```

11. 创建一个词列表并将其存储在变量 sent1 中,现在赋值 sent2=sent1,修改 sent1 中的一个项目,验证 sent2 改变了。

(1)现在尝试同样的练习,但使用 sent2=sent1[:]替换赋值语句,再次修改 sent1 看看 sent2 会发生什么,并解释为什么会这样。

```
sent1=['I', 'turned', 'off', 'the', 'spectroroute', '.']
sent2=sent1
sent1.append('update')
sent2
sent1=['I', 'turned', 'off', 'the', 'spectroroute', '.']
sent2=sent1[:]
sent1.append('update')
sent2
```

(2)现在定义 text1 为一个字符串列表的列表(如表示由多个句子组成的文本)。现在赋值 text2=text1[:],分配一个新值给其中一个词,例如 text1[1][1]='Monty'。检查该操作对 text2 做了什么,并解释为什么。

```
text1=[['I', 'turned', 'off', 'the', 'spectroroute', '.'], ['Where', 'are', 'we', 'going', 'after', 'this', '?']]
text2=text1[:]
text1[1][1]='Monty'
text2
```

(3)导入 Python 的 deepcopy()函数(即 from copy import deepcopy),查询其文档,使用它生成任一对象的新副本。

```
from copy import deepcopy
text1=[['I', 'turned', 'off', 'the', 'spectroroute', '.'], ['Where', 'are', 'we', 'going', 'after', 'this', '?']]
```

```
text2=deepcopy(text1)
text1[1][1]='Monty'
text2
```

英文题目：

Create a list of words and store it in a variable sent1. Now assign sent2 = sent1. Modify one of the items in sent1 and verify that sent2 has changed.Now try the same exercise but instead assign sent2 = sent1[:]. Modify sent1 again and see what happens to sent2. Explain.Now define text1 to be a list of lists of strings (e.g. to represent a text consisting of multiple sentences. Now assign text2 = text1[:], assign a new value to one of the words, e.g. text1[1][1] = 'Monty'. Check what this did to text2. Explain.Load Python's deepcopy() function (i.e. from copy import deepcopy), consult its documentation, and test that it makes a fresh copy of any object.

12. 使用列表乘法初始化 n×m 的空字符串列表的列表，例如 word_table=[['']*n]*m。当你设置其中一个值（例如 word_table[1][2]="hello"）时会发生什么事？解释为什么会出现这种情况。现在写一个表达式，使用 range()构造一个列表，表明它没有这个问题。

```
m, n=3, 4
word_table=[['']*n]*m
word_table[1][2]="hello"
[['', '', 'hello', ''], ['', '', 'hello', ''], ['', '', 'hello', '']]
m, n=3, 4
word_table=[[''  for j in range(n)] for i in range(m)]
word_table[1][2]="hello"
[['', '', '', ''], ['', '', 'hello', ''], ['', '', '', '']]
```

13. 写代码初始化一个名为 word_vowels 的二维数组的集合，处理一个词列表，添加每个词到 word_vowels[l][v]，其中 l 是词的长度，v 是它包含的元音的数量。

```
from pprint import pprint
def vow(w):
    return sum(1 for c in w.lower() if c in 'aeiou')

ws = ['The', 'Fulton', 'County', 'Grand', 'Jury', 'said', 'Friday', 'an', 'investigation', 'of', "Atlanta's", 'recent', 'primary', 'election', 'produced', '``', 'no', 'evidence', "'", 'that', 'any', 'irregularities', 'to', 'place', '.']
r=max(len(w) for w in ws)
```

```
c=max(vow(w) for w in ws)
word_vowels=[[[] for j in range(c)] for i in range(r)]
[word_vowels[len(w)][vow(w)].append(w) for w in ws]
pprint(word_vowels)
```

14. 写一个函数 novel10(text)输出所有在一个文本最后 10%出现而之前没有遇到过的词。

```
import nltk
def novel10(text):
    size=int(len(text)*0.9)
    training_sents, test_sents=text[:size], text[size:]
    return set(test_sents)-set(training_sents)

sents = ['The', 'Fulton', 'County', 'Grand', 'Jury', 'said', 'Friday',
'an', 'investigation', 'of', "Atlanta's", 'recent', 'primary', 'election',
'produced', '``', 'no', 'evidence', "''", 'that', 'any', 'irregularities',
'to', 'place', '.']
novel10(sents)      import nltk
def novel10(text):
    size=int(len(text)*0.9)
    training_text, test_text=text[:size], text[size:]
    [x for x in test_data if x not in training_text]

ws = ['The', 'Fulton', 'County', 'Grand', 'Jury', 'said', 'Friday', 'an',
'investigation', 'of', "Atlanta's", 'recent', 'primary', 'election',
'produced', '``', 'no', 'evidence', "''", 'that', 'any', 'irregularities',
'to', 'place', '.']
novel10(text)
```

15. 写一个程序，分割和计数一个句子中的词，句子为一个单独的字符串，输出每一个词和词的频率，每行一个，按字母顺序排列。

```
import nltk
sent='The Fulton County Grand Jury said Friday an investigation of'
sent=sent.split()
fd=nltk.FreqDist(sent)
for w in fd:
    print w, fd[w]
```

16. 阅读有关 Gematria 的内容。它是一种方法，分配一个数字给词汇，具有相同数字的词之间映射以发现文本隐藏的含义。

（1）写一个函数 gematria()，根据 letter_vals 中的字母值，累加一个词的字母的数值。

```
>>>letter_vals={'a':1, 'b':2, 'c':3, 'd':4, 'e':5, 'f':80, 'g':3,
'h':8, 'i':10, 'j':10, 'k':20, 'l':30, 'm':40, 'n':50, 'o':70, 'p':80,
'q':100, 'r':200, 's':300, 't':400, 'u':6, 'v':6, 'w':800, 'x':60,
'y':10, 'z':7}
   def gematria(letter_vals, w):
       return sum(letter_vals[c] for c in w.lower() if c.isalpha())

letter_vals={'a':1, 'b':2, 'c':3, 'd':4, 'e':5, 'f':80, 'g':3,
'h':8, 'i':10, 'j':10, 'k':20, 'l':30, 'm':40, 'n':50, 'o':70, 'p':80,
'q':100, 'r':200, 's':300, 't':400, 'u':6, 'v':6, 'w':800, 'x':60, '
y':10, 'z':7}
word='The'
gematria(letter_vals, word) #413
```

（2）处理一个语料库（如 nltk.corpus.state_union），对每个文档计数它有多少词的字母数值为 666。

```
import nltk
from nltk.corpus import state_union
def gematria(letter_vals, w):
    return sum(letter_vals[c] for c in w.lower() if c.isalpha())

letter_vals={'a':1, 'b':2, 'c':3, 'd':4, 'e':5, 'f':80, 'g':3,
'h':8, 'i':10, 'j':10, 'k':20, 'l':30, 'm':40, 'n':50, 'o':70, 'p':80,
'q':100, 'r':200, 's':300, 't':400, 'u':6, 'v':6, 'w':800, 'x':60,
'y':10, 'z':7}
ws=state_union.words()
sum(1 for w in ws if gematria(letter_vals, w)==666) #115
```

（3）写一个函数 decode() 来处理文本，随机替换词汇为它们的 Gematria 等值的词，以发现文本的"隐藏的含义"。

```
import nltk
import random
from nltk.corpus import state_union
letter_vals={'a':1, 'b':2, 'c':3, 'd':4, 'e':5, 'f':80, 'g':3,
'h':8, 'i':10, 'j':10, 'k':20, 'l':30, 'm':40, 'n':50, 'o':70, 'p':80,
'q':100, 'r':200, 's':300, 't':400, 'u':6, 'v':6, 'w':800, 'x':60,
'y':10, 'z':7}
state_union_words = list(state_union.words())
def gematria(letter_vals, w):
```

```
        return sum(letter_vals[c] for c in w.lower() if c.isalpha())

    def replace(w):
        for x in state_union_words:
            if gematria(letter_vals, w) == gematria(letter_vals, x):
                return x
        return w

    def decode(text):
        random.shuffle(state_union_words)
        return [replace(w) for w in text]

    text=['The', 'Fulton', 'County', 'Grand', 'Jury', 'said', 'Friday', 'an',
'investigation', 'of', "Atlanta's", 'recent', 'primary', 'election',
'produced', '``', 'no', 'evidence', "''", 'that', 'any', 'irregularities',
'to', 'place', '.']
    decode(text)
```

17. 写一个函数 shorten(text, n)处理文本, 省略文本中前 n 个最频繁出现的词。它的可读性会如何？

```
    import nltk
    import collections
    from nltk.corpus import brown
    from collections import Counter
    brown_words = brown.words(categories = 'news')
    brown_sents = brown.sents(categories = 'news')
    def shorten(text, n):
        fd = nltk.FreqDist(brown_words)
        mc = Counter(fd).most_common(n)
        mc = [x for x, y in mc]
        return [w for w in text if w not in mc]

    text = brown_sents[0]
    text
    n = 20
    shorten(text, n)
    ['The', 'Fulton', 'County', 'Grand', 'Jury', 'said', 'Friday', 'an',
'investigation', 'of', "Atlanta's", 'recent', 'primary', 'election',
'produced',
```

```
'``', 'no', 'evidence', "'", 'that', 'any', 'irregularities', 'to', 'place', '.']
['Fulton', 'County', 'Grand', 'Jury', 'said', 'Friday', 'an', 'investigation',
"Atlanta's", 'recent', 'primary', 'election', 'produced', 'no', 'evidence',
'any', 'irregularities', 'to', 'place']
```

可读性好。

18. 写一段代码输出词汇的索引，允许别人根据其含义查找词汇（或它们的发音，词汇条目中包含的任何属性都行）。

19. 写一个列表推导排序 WordNet 中与给定同义词集接近的同义词集的列表。例如，给定同义词集 minke_whale.n.01、orca.n.01、novel.n.01 和 tortoise.n.01，按照它们与 right_whale.n.01 的 path_distance() 对它们进行排序。

20. 写一个函数处理一个词列表（含重复项），返回一个按照频率递减排序的词列表（没有重复项）。例如，如果输入列表中包含词 table 的 10 个实例，chair 的 9 个实例，那么在输出列表中 table 会出现在 chair 前面。

```
import nltk
wordlist=["table", "table", "table", "table", "table", "table",
"table", "table", "table", "table", "chair", "chair", "chair", "chair",
"chair", "chair", "chair", "chair", "chair"]
list(nltk.FreqDist(wordlist))
```

21. 写一个函数，以一个文本和一个词汇表作为它的参数，返回在文本中出现但不在词汇表中的一组词。两个参数都可以表示为字符串列表。你能使用 set.difference() 在一行代码中做这些吗？

```
def f(text, lexicon):
    return set(text).difference(set(lexicon))
text=['As', 'the', 'first', 'generation', 'born', 'under', 'the',
'one-child', 'policy']
lexicon=['the', 'family', 'environment', 'of', 'the', '1980s',
'generation', 'is']
f(text, lexicon)
```

22. 从 Python 的标准库 operator 模块导入 itemgetter() 函数（即 from operator import itemgetter）。创建一个包含几个词的列表 words。现在尝试调用：sorted(words, key=itemgetter(1)) 和 sorted(words, key=itemgetter(-1))。解释 itemgetter() 正在做什么。

```
from operator import itemgetter
words=["i", "love", "you"]
sorted(words, key=itemgetter(1))
sorted(words, key=itemgetter(-1)) #?
sorted(words, key=itemgetter(0)) #?
```

获取对象的第几个值。words 为列表，其元素为字符串，可以看出是字符串或字符串。key=itemgetter(1)指定按元素第 2 个域（下标为 1 的域，这里越界）排序。key=itemgetter(-1)指定按元素第 1 个域（下标为 0 的域，即字符串）排序。

key=itemgetter(0)是否是指定按元素第 1 个域（下标为 0 的域，即字符串）排序？key=itemgetter(-1)与 key=itemgetter(-1)一样吗？

23. 写一个递归函数 loup(trie, key)，在查找树中查找一个关键字，返回找到的值。扩展这个函数，当一个词由其前缀唯一确定时，返回这个词。例如，vanguard 是以 vang-开头的唯一的词，所以 loup(trie, 'vang')应该返回与 loup(trie, 'vanguard')相同的内容。

```
def loup(trie, key):
    if key:
        first, rest=key[0], key[1:]
        return loup(trie[first], rest)
    else:
        return trie['value']
def extend(trie):
    if 'value' not in trie:
        first=trie.keys()[0]
        return extend(trie[first])
    else:
        return trie['value']
```

```
def loup(trie, key):
    if key:
        first, rest=key[0], key[1:]
        return loup(trie[first], rest)
    else:
        return extend(trie)
```

24. 阅读关于关键字联动的内容。从 NLTK 的莎士比亚语料库中提取关键字，使用 NetworkX 包，画出关键字联动网络。

25. 阅读有关字符串编辑距离和 Levenshtein 算法的内容，尝试使用 nltk.edit_dist()。它使用的是动态规划的何种方式？是自下而上或自上而下的方法吗？

26. Catalan 数出现在组合数学的许多应用中，包括解析树的计数。该级数定义如下：$C_0=1$，$C_{n+1}=\sum_{0 \cdots n}(C_i C_{n-i})$。

（1）编写一个递归函数计算第 n 个 Catalan 数 C_n。

```
def C(n):
    if n==0:
        return 1
```

```
    else:
        return sum(C(i)*C(n-1-i) for i in range(n))
```

(2)现在写另一个函数使用动态规划做这个计算。

```
C=[0 for i in range(26)]
C[0]=1

def f(n):
    if C[n]>0:
        return C[n]
    else:
        C[n]=sum(f(i)*f(n-1-i) for i in range(n))
        return C[n]
```

(3)使用 timeit 模块比较当 n 增加时这些函数的性能。

```
from timeit import repeat
def C(n):
    if n==0:
        return 1
    else:
        return sum(C(i)*C(n-1-i) for i in range(n))

for i in range(10):
    t=repeat('C(%d)' % i, 'from __main__ import C', number=1000)
    print min(t)
from timeit import repeat
C=[0 for i in range(26)]
C[0]=1
def f(n):
    if C[n]>0:
        return C[n]
    else:
        C[n]=sum(f(i)*f(n-1-i) for i in range(n))
        return C[n]

for i in range(26):
    t=repeat('f(%d)' % i, 'from __main__ import f', number=1000)
    print min(t)
from timeit import repeat
```

```
def C(n):
    if n==0:
        return 1
    else:
        return sum(C(i)*C(n-1-i) for i in range(n))

t=repeat('C(3)', 'from __main__ import C', repeat=3, number=1)
print t
```

27. 写一个递归函数，按字母顺序排列输出一个查找树。

```
chair:'flesh'
---t:'cat'
--ic:'stylish'
---en:'dog'
def insert(trie, key, value):
    if key:
        first, rest=key[0], key[1:]
        if first not in trie:
            trie[first]={}
        insert(trie[first], rest, value)
    else:
        trie['value']=value

def pp(trie, x):
    keys=trie.keys()
    values=trie.values()
    if keys[0]=='value':
        print x+':'+repr(values[0])
        return
    else:
        i=0
        for j in keys:
            x+=j
            pp(values[i], x)
            i+=1
            x='-'*len(x[:-1])

import nltk
trie=nltk.defaultdict(dict)
```

```
insert(trie, 'chat', 'cat')
insert(trie, 'chien', 'dog')
insert(trie, 'chair', 'flesh')
insert(trie, 'chic', 'stylish')

x=''
pp(trie, x)
```

28. 在查找树数据结构的帮助下，编写一个递归函数处理文本，定位在每个词的独特点，并丢弃每个词的其余部分。这样压缩了多少？产生的文本可读性如何？

```
import nltk
def insert(trie, key, value):
    if key:
        first, rest=key[0], key[1:]
        if first not in trie:
            trie[first]={}
        insert(trie[first], rest, value)
    else:
        trie['value']=value

def dfs(trie):
    j=0
    for k in trie.keys():
        if k=='value':
            j+=1
        else:
            j+=dfs(trie[k])
    return j

def locate(trie, key):
    if dfs(trie)==1:
        return key
    else:
        if key:
            first, rest=key[0], key[1:]
            return locate(trie[first], rest)
        else:
            return ''
```

```
trie=nltk.defaultdict(dict)
insert(trie, 'cat', 'cat')
insert(trie, 'catch', 'catch')
insert(trie, 'vanguard', 'vanguard')
x=locate(trie, 'catch')
'catch'.rstrip(x)
```

29. 以一个单独的长字符串的形式，获取一些原始文本。使用 Python 的 textwrap 模块将它分割成多行。现在，写代码在词之间添加额外的空格，以调整输出。每一行必须具有相同的宽度，每行均匀分布（行不能以空白开始或结束）。

```
'''
格式化:从列表到字符串
文本换行
背景:textwrap 的不足分析
类似于Word两端对齐(分散对齐)
'''
from textwrap import fill
saying=['After', 'all', 'is', 'said', 'and', 'done', ',', 'more',
'is', 'said', 'than', 'done', '.', 'After', 'all', 'is', 'said', 'and',
'done', ',', 'more', 'is', 'said', 'than', 'done', '.']
format='%s_(%d), '
pieces=[format % (word, len(word)) for word in saying]
output=' '.join(pieces)
wrapped=fill(output)
wrapped=wrapped.split('\n')
len(wrapped)
[len(s) for s in wrapped]
print wrapped.replace('_', ' ')  #这里有一个空格!
'''
'After_(5), all_(3), is_(2), said_(4), and_(3), done_(4), ,_(1),
\n
more_(4), is_(2), said_(4), than_(4), done_(4), ._(1), '
'''
```

30. 开发一个简单的挖掘总结工具，输出一个文档中包含最高总词频的句子。使用 FreqDist() 计数词频，并使用 sum 累加每个句子中词的频率。按照句子的得分排序。最后，输出文档中 n 个最高得分的句子。仔细检查程序的设计，尤其是关于这种双重排序的方法，确保程序写得简洁、易读。

```
import nltk
raw='This is a bo . She sells sea shell by the sea shore .'
words=nltk.word_tenize(raw)
fd=nltk.FreqDist(words)
sents=nltk.sent_tenize(raw)
text=[nltk.word_tenize(sent) for sent in sents]
a=[sum(fd[w] for w in sent) for sent in text]
b=[(f, i) for i, f in enumerate(a)]
c=sorted(b, reverse=True)
d=[i for _, i in c][:2]
[text[i] for i in d]
```

31. 使用 Network X 包可视化一个形容词的网络，其中的边表示相同和不同的语义倾向。

32. 设计一个算法找出一个文档集合中统计学上不可能的短语。

将概率小于 0.05 看作是小概率事件，小概率事件在一次试验中几乎不可能发生。

33. 写一个程序实现发现一种 n×n 的四方联词的蛮力算法：纵横字谜，它的第 n 行的词与第 n 列的词相同（回溯法可解此题）。

阅读材料

Python 音频处理

（1）读取本地音频数据。

处理音频的第一步是需要让计算机"听到"声音，这里使用 Python 标准库中自带的 wave 模块进行音频参数的获取。

① 导入 wave 模块。

② 使用 wave 中的函数 open 打开音频文件，wave.open(file，mode)函数带有两个参数，第一个参数 file 是所需要打开的文件名及路径，使用字符串表示；第二个参数 mode 是打开的模式，用字符串表示'rb'或'wb'。

③ 打开音频后使用 getparams()获取音频基本的相关参数。nchannels：声道数；sampwidth：量化位数或量化深度；framerate：采样频率；nframes：采样点数。

```
#导入 wave 模块
import wave
#用于绘制波形图
import matplotlib.pyplot as plt
#用于计算波形数据
import numpy as np
#用于系统处理，如读取本地音频文件
import os
#打开 WAV 文档
f=wave.open(r"2.wav", 'rb')
#读取格式信息
params=f.getparams()
nchannels, sampwidth, framerate, nframes=params[:4]
print(framerate)
```

（2）读取单通道音频并绘制波形图。

① 通过第 1 步，可以继续读取音频数据本身保存为字符串格式。

readframes：读取声音数据，传递一个参数指定需要读取的长度，以取样点为单位，返回的是二进制数据。在 Python 中用字符串表示二进制数据。

```
strData=f.readframes(nframes)
```

② 如果需要绘制波形图，则将字符串格式的音频数据转化为 int 类型

frombuffer：根据声道数和量化单位，将读取的二进制数据转换为一个可以计算的数组。通过 frombuffer 函数将二进制转换为整型数组，通过其参数 dtype 指定转换后的数据格式。

```
waveData=np.frombuffer(strData, dtype=np.int16)
```

此处需要使用 numpy 进行数据格式的转化。

③ 将幅值归一化。

把数据变成范围在(0,1)的小数,主要是为了数据处理后方便提取,把数据映射到 0~1 处理,更加快捷。

```
waveData=waveData*1.0/max(abs(waveData))
```

不进行此步也可以画出波形图,可以尝试不用此步,找出画出的波形图的不同。

④ 绘制图像。

通过取样点数和取样频率计算出取样的时间。

```
time=np.arange(0, nframes)*(1.0/framerate)
```

文件大小为 96 kB,时长为 00:02,声道为单声道,音质为 HQ(高品质),比特率为 352 kb/s,采样率为 22 050 k。

```
import wave
#导入 wave 模块
import matplotlib.pyplot as plt
#用于绘制波形图
import numpy as np
#用于计算波形数据
import os
#用于系统处理,如读取本地音频文件
f=wave.open(r"di.wav", 'rb')
params=f.getparams()
nchannels, sampwidth, framerate, nframes=params[:4]
print(framerate)
#读取波形数据
strData=f.readframes(nframes)
#将字符串转换为 16 位整数
waveData=np.frombuffer(strData, dtype=np.int16)
#幅值归一化
waveData=waveData*1.0/max(abs(waveData))
#计算音频的时间
time=np.arange(0, nframes)*1.0/framerate
plt.plot(time, waveData)
plt.xlabel("Time(s)")
plt.ylabel("Amplitude")
plt.title("Single channel wavedata")
plt.show()
```

波形图效果如图 4-3 所示。

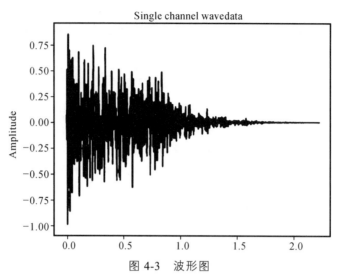

图 4-3 波形图

```
>>> f
<wave.Wave_read object at 0x000000F33993B198>
>>> params
_wave_params(nchannels=1, sampwidth=2, framerate=22050, nframes=49208, comptype=
'NONE', compname='not compressed')
>>> time
array([0.00000000e+00, 4.53514739e-05, 9.07029478e-05, ..., 2.23160998e+00])
 \x00\x02\x00\x02\x00\x02\x00\x00\x00\x00\x00\xf9\xff\xf9\xff\xff\xf9\xff\xf8\xff'
>>> waveData
array([-0.17182522, -0.14838416, -0.05029586, ..., -0.00031862, -0.00036413])
```

第 5 章 词性标注

5.1 词性标注器

词性标注器（Parts-Of-Speech Tagger，POS Tagger）处理一个词序列，为每个词附加一个词性标记。

词序列的词性标注。

```
>>> from nltk import word_tokenize, pos_tag
>>> text=word_tokenize("And now for something completely different")
>>> pos_tag(text)
resource_val = pickle.load(opened_resource, encoding='iso-8859-1')
[('And', 'CC'), ('now', 'RB'), ('for', 'IN'), ('something', 'NN'),
('completely', 'RB'), ('different', 'JJ')]
```

And 是 CC（并列连词），now 和 completely 是 RB（副词），for 是 IN（介词），something 是 NN（名词），different 是 JJ（形容词），标注结果正确。

同形异音异义词的词性标注。

```
>>> from nltk import word_tokenize, pos_tag
>>> text=word_tokenize("They refuse to permit us to obtain the refuse permit.")
>>> pos_tag(text)
[('They', 'PRP'), ('refuse', 'VBP'), ('to', 'TO'), ('permit', 'VB'),
('us', 'PRP'), ('to', 'TO'), ('obtain', 'VB'), ('the', 'DT'), ('refuse',
'NN'), ('permit', 'NN'), ('.', '.')]
```

refuse 和 permit 都以一般现在时动词（VBP）和名词（NN）形式出现，refUSE 是一个动词，意为"拒绝"，而 REFuse 是一个名词，意思是"垃圾"，即它们不是同音词。这些类别中很多都源于对文本中词分布浅层的分析，涉及 woman、bought、over 和 the。text.similar(w) 方法为一个词 w 找出所有上下文 $w_1 w w_2$ (3-gram)，然后找出所有出现在相同上下文中的词 w'，即 $w_1 w' w_2$。

文本中词分布分析。

```
>>> from nltk.text import Text
>>> from nltk.corpus import brown
>>> text=Text(w.lower() for w in brown.words())
>>> text.similar('woman')
man day time year moment car world family house boy country child
job state war place way girl word case
>>> text.similar('bought')
made put said done seen found had left given heard set was brought
got been felt took that called told
>>> text.similar('over')
in on to of and for with from at by that into as up out down through
all is about
>>> text.similar('the')
a his this their its her an that our any all one these my in your
no some other and
```

搜索 woman 找到的是名词，搜索 bought 找到的大部分是动词，搜索 over 一般会找到介词，搜索 the 找到几个限定词，于是，将 woman 标注为名词，将 bought 标注为动词，将 over 标注为介词，将 the 标注为限定词。

词标记的识别。

```
>>> from nltk import word_tokenize, pos_tag
>>> #text=['This', 'is', 'a', 'a', 'book', '.']
... #text="This is a book.".split()
... text=word_tokenize("This is a book.")
>>> pos_tag(text)
[('This', 'DT'), ('is', 'VBZ'), ('a', 'DT'), ('book', 'NN'), ('.', '.')]
```

'This'是定冠词，'is'是动词第三人称单数，'a'是定冠词，'book'是名词，'.'是标点符号，识别结果正确。

5.2 标注语料库

5.2.1 表示已标注的标识符

按照 NLTK 的约定,一个已标注的标识符使用一个由标识符和标记组成的二元组来表示。可以使用函数 str2tuple()从表示一个已标注的标识符的标准字符串创建一个特殊元组。

```
from nltk.tag import str2tuple
tagged_ten=str2tuple('fly/NN')
tagged_ten #('fly', 'NN')
tagged_ten[0] #'fly'
tagged_ten[1] #'NN'
tagged_ten=('fly', 'NN')
tagged_ten='fly', 'NN'
tagged_ten=tuple(['fly', 'NN'])
tagged_ten=tuple({'fly', 'NN'})
```

可以直接从一个字符串构造一个已标注的标识符的列表。第一步是对字符串分词,以便能访问单独的词/标记字符串,然后转换成一个个二元组(使用 str2tuple())。

```
from nltk.tag import str2tuple
from nltk import word_tenize
sent='''This/DT is/VBZ a/DT bo/NN ./.'''
[str2tuple(t) for t in sent.split()]
[str2tuple(t) for t in word_tenize(sent)]
```

5.2.2 已标注语料库的读取

NLTK 中若干语料库已标注了词性,各种语料库使用各种格式存储词性标记。NLTK 中的语料库阅读器提供了一个统一的接口,可以处理不同的文件格式。brown 语料库的阅读器按如下方式表示数据。

```
Python 2.7.11 |Anaconda 2.5.0 (64-bit)| (default, Jan 29 2016,
14:26:21) [MSC v.1500 64 bit (AMD64)] on win32
Type "help", "copyright", "credits" or "license" for more information.
Anaconda is brought to you by Continuum Analytics.
Please check out: http://continuum.io/thanks and https://anaconda.org
>>> from nltk.corpus import brown
>>> brown.tagged_words()
[(u'The', u'AT'), (u'Fulton', u'NP-TL'), ...]
>>> brown.tagged_words(tagset='universal')
[(u'The', u'DET'), (u'Fulton', u'NOUN'), ...]
>>> brown.words()
[u'The', u'Fulton', u'County', u'Grand', u'Jury', ...]
>>> from nltk.tag import str2tuple
>>> str2tuple('The/AT')
('The', 'AT')
```

只要语料库包含已标注的文本，NLTK 语料库接口都有一个 tagged_words()方法。并非所有的语料库都采用同一组标记。最初，想避免这些标记集的复杂化，所以使用一个内置的到一个简化的标记集的映射。

```
Python 2.7.11 |Anaconda 2.5.0 (64-bit)| (default, Jan 29 2016,
14:26:21) [MSC v.1500 64 bit (AMD64)] on win32
Type "help", "copyright", "credits" or "license" for more information.
Anaconda is brought to you by Continuum Analytics.
Please check out: http://continuum.io/thanks and https://anaconda.org
>>> from nltk.tag import pos_tag
>>> from nltk.tokenize import word_tokenize
>>> # Default Penntreebank tagset.
... pos_tag(word_tokenize("John's big idea isn't all that bad."))
[('John', 'NNP'), ("'s", 'POS'), ('big', 'JJ'), ('idea', 'NN'),
('is', 'VBZ'), ("n't", 'RB'), ('all', 'PDT'), ('that', 'DT'), ('bad',
'JJ'), ('.', '.')]
>>> # Universal POS tags.
... pos_tag(word_tokenize("John's big idea isn't all that bad."),
tagset='universal')
[('John', u'NOUN'), ("'s", u'PRT'), ('big', u'ADJ'), ('idea',
u'NOUN'), ('is', u'VERB'), ("n't", u'ADV'), ('all', u'DET'), ('that',
u'DET'), ('bad', u'ADJ'), ('.', u'.')]
>>> from nltk.corpus import brown
>>> brown.tagged_words(tagset='universal')
[(u'The', u'DET'), (u'Fulton', u'NOUN'), ...]
>>> brown.tagged_words()
[(u'The', u'AT'), (u'Fulton', u'NP-TL'), ...]
```

如果语料库也被分割成句子，tagged_sents()方法可以将已标注的词划分成句子，而不是将它们表示成一个大列表。

5.2.3　简化的词性标记集

已标注的语料库使用许多不同的标记集约定来标注词汇。

让我们来看看这些标记中哪些是 brown 语料库的 news 类中最常见的。

```
from nltk.corpus import brown
from nltk import FreqDist
brown_news_tagged=brown.tagged_words(categories='news',
tagset='universal')
tag_fd=FreqDist(tag for _, tag in brown_news_tagged)
tag_fd.keys()
```

使用 tag_fd.plot(cumulative=True)为上面显示的频率分布绘图，如图 5-1 所示。标注为上述列表中的前 5 个标记的词的比例是多少？

```
import nltk
from nltk.corpus import brown
brown_news_tagged=brown.tagged_words(categories='news',
simplify_tags=True)
tag_fd=nltk.FreqDist(tag for word, tag in brown_news_tagged)
tag_fd.keys()
tag_fd.plot(cumulative=True)
```

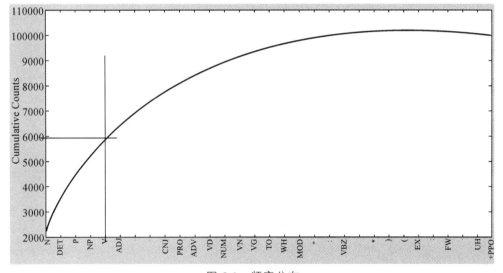

图 5-1 频率分布

```
['N', 'DET', 'P', 'NP', 'V', 'ADJ', ',', '.', 'CNJ', 'PRO', 'ADV',
'VD', 'NUM', 'VN', 'VG', 'TO', 'WH', 'MOD', '``', "''", 'VBZ', '', '*',
')', '(', 'EX', ':', 'FW', "'", 'UH', 'VB+PPO']
import nltk
from nltk.corpus import brown
brown_news_tagged=brown.tagged_words(categories='news',
simplify_tags=True)
tag_fd=nltk.FreqDist(tag for word, tag in brown_news_tagged)
```

```
cumulative=0.0
for word in tag_fd.keys()[:5]:
    cumulative+=tag_fd[word]*100.0/tag_fd.N()

print '%.1f%%' % cumulative #59.1%
import nltk
from nltk.corpus import brown
brown_news_tagged=brown.tagged_words(categories='news', simplify_tags=True)
tag_fd=nltk.FreqDist(tag for word, tag in brown_news_tagged)
cumulative=0.0
for word in tag_fd.keys()[:5]:
    cumulative+=tag_fd.freq(word)

print '%.1f%%' % (cumulative*100) #59.1%
```

可以使用这些标记结合一个图形化的 POS 一致性工具进行搜索。

```
import nltk
nltk.app.concordance()
```

用它来寻找任一词和 POS 标记的组合，如 N N N N，hit/VD，hit/VN 或 the ADJ man，如图 5-2 所示。

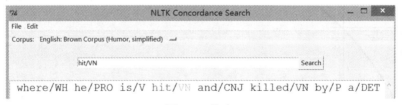

图 5-2 搜索

5.2.4 名 词

名词可能出现在限定词和形容词之后<DT><ADJ><NOUN>，可以是动词的主语或宾语，见表 5-1。

表 5-1 一些名词的句法模式

词	限定词之后	动词的主语
woman	the woman who I saw yesterday	the woman sat down
Scotland	the Scotland I remember as a child	Scotland has five million people
bo	the bo I bought yesterday	this bo recounts the colonization of Australia
intelligence	the intelligence displayed by the child	Mary's intelligence impressed her teachers

简化的名词标记对普通名词是 NOUN，专有名词是 NP。让我们检查一些已标注的文本，看看哪些词类出现在一个名词前，频率最高的在最前面。首先构建一个双连词列表，它的成员是它们自己的词-标记对，例如，[(('The', 'DET'), ('Fulton', 'NOUN')), (('Fulton', 'NP'), ('County', 'N'))]。然后构建一个双连词标记部分的 FreqDist。

```
from nltk.corpus import brown
from nltk import bigrams, FreqDist
brown_news_tagged=brown.tagged_words(categories='news',
tagset='universal')
word_tag_pairs=bigrams(brown_news_tagged)
list(FreqDist(a[1] for (a, b) in word_tag_pairs if b[1]=='NOUN'))
brown_news_tagged[:3]
list(word_tag_pairs)[:3]
(  (u'The', u'DET'),  (u'Fulton', u'NOUN')  )
(      a      ,      b      )
```

这证实了名词出现在限定词和形容词之后，包括数字形容词、数词，标注为 NUM。

5.2.5 动　词

动词是用来描述事件和行动的词，如 fall 和 eat，见表 5-2。

表 5-2　一些动词的句法模式

词	例子	修饰符与修饰语
fall	Rome fell	Dot com stocks suddenly fell like a stone
eat	Mice eat cheese	John ate the pizza with gusto

新闻文本中最常见的动词是什么？让我们按频率排序所有动词。

```
from nltk.corpus import treebank
from nltk import FreqDist
wsj=treebank.tagged_words()
word_tag_fd=FreqDist(wsj)
[word+"/"+tag for (word, tag) in word_tag_fd if tag.startswith('V')][:3]
```

注意：频率分布中计算的项目是词-标记对。由于词汇和标记是成对的，可以把词作为条件，标记作为事件，使用条件-事件对的列表初始化一个条件频率分布。这让我们看到了一个给定的词的标记的频率顺序列表。

```
import nltk
wsj=nltk.corpus.treebank.tagged_words(simplify_tags=True)
cfd1=nltk.ConditionalFreqDist(wsj)
```

```
cfd1['yield'].keys() #['V', 'N']
cfd1['cut'].keys() #['V', 'VD', 'N', 'VN']
len(cfd1['yield']) #2
import nltk
text='Excuse me . Is there any summer resort around the town ?'
text=nltk.word_tenize(text)
text=nltk.pos_tag(text)
wsj=text
word_tag_fd=nltk.FreqDist(wsj)
[word+"/"+tag for (word, tag) in word_tag_fd if tag.startswith('V')]
#['Is/VBZ']
cfd1=nltk.ConditionalFreqDist(wsj)
cfd1['Excuse'].keys() #['NNP']
cfd1['me'].keys() #['PRP']
```

可以颠倒配对的顺序，标记作为条件，词汇作为事件。现在可以看到对于一个给定的标记可能的词。

```
import nltk
wsj=nltk.corpus.treebank.tagged_words(simplify_tags=True)
cfd2=nltk.ConditionalFreqDist((tag, word) for word, tag in wsj)
cfd2['VN'].keys()
```

要弄清 VD（过去式）和 VN（过去分词）之间的区别，先找到可以同是 VD 和 VN 的词汇，看看它们周围的一些文字。

```
import nltk
wsj=nltk.corpus.treebank.tagged_words(simplify_tags=True)
idx1=wsj.index(('kicked', 'VD'))
wsj[idx1-4:idx1+1] #[('While', 'P'), ('program', 'N'), ('trades',
'N'), ('swiftly', 'ADV'), ('kicked', 'VD')]
idx2=wsj.index(('kicked', 'VN'))
wsj[idx2-4:idx2+1] #[('head', 'N'), ('of', 'P'), ('state', 'N'),
('has', 'V'), ('kicked', 'VN')]
```

在这种情况下，可以看到，过去分词 kicked 前面是助动词 have 的形式。这是普遍、真实的吗？

通过 cfd2['VN'].keys() 指定一个过去分词的列表，尝试收集所有直接在列表中项目前面的词-标记对。

```
import nltk
wsj=nltk.corpus.treebank.tagged_words(simplify_tags=True)
cfd2=nltk.ConditionalFreqDist((tag, word) for word, tag in wsj)
```

```
for word in cfd2['VN'].keys()[:3]:
    idx3=wsj.index((word, 'VN'))
    wsj[idx3-1:idx3+1]
[('had', 'VD'), ('been', 'VN')]
[('the', 'DET'), ('expected', 'VN')]
[('collections', 'N'), ('made', 'VN')]
```

5.2.6 未简化的标记

下面的程序找出所有以 NN 开始的标记，并为每个标记提供了几个示例词汇。会看到有许多名词的变种，含有$的是名词所有格，含有 S 的是复数名词以及含有 P 的专有名词。此外，大多数标记都有后缀修饰符，-NC 表示引用，-HL 表示标题中的词，-TL 表示标题。

查找出所有以 NN 开始的标记，并为每个标记提供了几个最频繁的示例词汇。

```
>>> from nltk import ConditionalFreqDist
>>> from nltk.corpus import brown
>>> from collections import Counter
>>> def findtags(tag_prefix, tagged_text):
...     cfd=ConditionalFreqDist((tag, word)
...     for word, tag in tagged_text if tag.startswith(tag_prefix))
...     return dict((tag, Counter(cfd[tag]).most_common(5)) for tag in cfd)
...
>>> tagdict=findtags('NN', brown.tagged_words(categories='news'))
>>> for tag in sorted(tagdict):
...     print tag, [x[0] for x in tagdict[tag]]
...
NN [u'year', u'time', u'state', u'week', u'man']
NN$ [u"year's", u"world's", u"state's", u"company's", u"nation's"]
NN$-HL [u"Golf's", u"Navy's"]
NN$-TL [u"President's", u"University's", u"Gallery's", u"Army's", u"Administration's"]
NN-HL [u'cut', u'condition', u'Salary', u'Question', u'business']
NN-NC [u'eva', u'ova', u'aya']
NN-TL [u'President', u'House', u'State', u'University', u'City']
NN-TL-HL [u'Fort', u'City', u'Commissioner', u'Grove', u'House']
NNS [u'years', u'members', u'people', u'sales', u'men']
NNS$ [u"children's", u"women's", u"janitors'", u"men's", u"taxpayers'"]
```

```
    NNS$-HL [u"Dealers'", u"Idols'"]
    NNS$-TL [u"Women's", u"States'", u"Giants'", u"Officers'", u"Bombers'"]
    NNS-HL [u'addresses', u'years', u'schools', u'dances', u'Creations']
    NNS-TL [u'States', u'Nations', u'Masters', u'Rules', u'Communists']
    NNS-TL-HL [u'Nations']

    import nltk
    text='Excuse me . Is there any summer resort around the town ?'
    text=nltk.word_tenize(text)
    text=nltk.pos_tag(text)
    tag_prefix, tagged_text='NN', text
    cfd=nltk.ConditionalFreqDist((tag, word) for word, tag in tagged_text if tag.startswith(tag_prefix))
    tagdict=dict((tag, cfd[tag].keys()[:5]) for tag in cfd.conditions())
    for tag in sorted(tagdict):
        print tag, tagdict[tag]
```

5.2.7 已标注语料库探索

假设我们正在研究词 often，想看看它是如何在文本中使用的，可以试着看看跟在 often 后面的词汇。

```
>>> from nltk import bigrams
>>> from nltk.corpus import brown
>>> brown_learned_text=brown.words(categories='learned')
>>> sorted(set(b for a, b in bigrams(brown_learned_text) if a=='often'))
[u',', u'.', u'accomplished', u'analytically', u'appear', u'apt', u'associated', u'assuming', u'became', u'become', u'been', u'began', u'call', u'called', u'carefully', u'chose', u'classified', u'colorful', u'composed', u'contain', u'differed', u'difficult', u'encountered', u'enough', u'equate', u'extremely', u'found', u'happens', u'have', u'ignored', u'in', u'involved', u'more', u'nee
>>> from nltk.text import Text
>>> Text(brown_learned_text).concordance('often')
Displaying 25 of 70 matches:
 r as high temperature research tools often more than 50% of the total energy in
```

```
ich is usually inorganic and is very often extremely finely divided
so as to ex
 ; and natural waters , in addition , often contain impurities such
as calcium s
```

然而，使用 tagged_words()方法查看跟随词的词性标记可能更有指导性。

```
Python 2.7.11 |Anaconda 2.5.0 (64-bit)| (default, Jan 29 2016,
14:26:21) [MSC v.1500 64 bit (AMD64)] on win32
>>> from nltk import bigrams, FreqDist
>>> from nltk.corpus import brown
>>> brown_lrnd_tagged=brown.tagged_words(categories='learned')
>>> tags=[b[1] for a, b in bigrams(brown_lrnd_tagged) if
a[0]=='often']
>>> fd=FreqDist(tags)
>>> fd.tabulate()
 VBN  VB  VBD  JJ  IN  RB  ,  CS  QL  WRB  TO
  15  10   8   5   4   3  3   3   3   1   1
```

注意：often 后面最高频率的词性是动词。名词从来没有在这个位置出现（在这个特别的语料中）。

```
import nltk
text='Excuse me . Is there any summer resort around the town ?'
text=nltk.word_tenize(text)
brown_lrnd_tagged=nltk.pos_tag(text)
tags=[b[1] for (a, b) in nltk.ibigrams(brown_lrnd_tagged) if
a[0]=='often']
fd=nltk.FreqDist(tags)
fd.tabulate()
```

接下来，让我们看一些较大范围的上下文，找出涉及特定标记和词序列的词（"<Verb>到<Verb>"）。考虑句子中的每个三词窗口，检查它们是否符合标准。如果标记匹配，则输出对应的词。

使用 POS 标记寻找三词短语。

```
import nltk
from nltk.corpus import brown
def process(sentence):
    for ((w1, t1), (w2, t2), (w3, t3)) in nltk.trigrams(sentence):
        if(t1.startswith('V') and t2=='TO' and t3.startswith('V')):
            print w1, w2, w3
```

```
for tagged_sent in brown.tagged_sents()[:20]:
    process(tagged_sent)
combined to achieve
continue to place
import nltk
cp=nltk.RegexpParser('CHUNK:{<V.*><TO><V.*>}')
brown=nltk.corpus.brown
for sent in brown.tagged_sents():
    tree=cp.parse(sent)
    for subtree in tree.subtrees():
        if subtree.node=='CHUNK':
            print subtree
import nltk
def process(sentence):
    for (w1, t1), (w2, t2), (w3, t3) in nltk.trigrams(sentence):
        if(t1.startswith('V') and t2=='RB' and t3.startswith('DT')):
            print w1, w2, w3

text='Excuse me . Is there any summer resort around the town ?'
text=nltk.word_tenize(text)
text=nltk.pos_tag(text)
process(text)
```

最后,看看与它们的标记关系高度模糊不清的词(有 3 种以上词性的词)。标注这样的词是因为它们各自的上下文可以帮助我们弄清楚标记之间的区别。

```
brown_news_tagged=brown.tagged_words(categories='news', simplify_tags=True)
    data=nltk.ConditionalFreqDist((word.lower(), tag) for (word, tag) in brown_news_tagged)
    for word in data.conditions():
        if len(data[word])>3:
            tags=data[word].keys()
            print word, ' '.join(tags)


best ADJ ADV NP V
better ADJ ADV V DET
close ADV ADJ V N
cut V N VN VD
```

```
even ADV DET ADJ V
hit V VD VN N
lay ADJ V NP VD
left VD ADJ N VN
like CNJ V ADJ P
near P ADJ ADV DET
open ADJ V N ADV
past N ADJ DET P
present ADJ ADV N V
read V VN VD NP
right ADJ N DET ADV
second NUM ADV DET N
text='Excuse me . Is there any summer resort around the town ?'
text=nltk.word_tenize(text)
text=nltk.pos_tag(text)
brown_news_tagged=text
data=nltk.ConditionalFreqDist((word.lower(), tag) for (word, tag) in brown_news_tagged)
for word in data.conditions():
    if len(data[word])>0:
        tags=data[word].keys()
        print word, ' '.join(tags)

? .
any DT
around IN
excuse NNP
is VBZ
me PRP
resort NN
summer NN
the DT
there RB
town NN
```

打开 POS 一致性工具 nltk.app.concordance()(nltk.text.Text.concordance)和加载完整的布朗语料库（简化标记集 simplify_tags=True）。现在挑选一些上面例子末尾处列出的词，看看词的标记如何与词的上下文相关。例如，搜索 near 会看到所有混合在一起的形式，搜索 near/ADJ 会看到它作为形容词使用，near N 会看到只是名词跟在后面的情况，等等。

5.3 使用 Python 字典映射词及其属性

(word, tag)二元组是词和词性标记的关联,一旦开始做词性标注,将会创建分配一个标记给一个词的程序。标记是在给定上下文中最可能的标记,可以认为这个过程是从词到标记的映射 tag=f(word)。在 Python 中映射是使用字典数据类型,在 PHP 中则称为关联数组或哈希数组。

5.3.1 索引列表和字典

我们已经看到,文本在 Python 中被视为一个词列表,列表的一个重要的属性是可以通过给出其索引来取特定项目,如 text1[100]。注意如何指定一个数字,然后取回一个词,一个整数索引帮助我们访问 Python 列表的内容,对比这种情况与频率分布,在那里指定一个词然后取回一个数字,如 fdist['monstrous']。它告诉我们一个给定的词在文本中出现的次数,用词查询类似于使用一本字典。

使用一个关键字,如某人的名字、一个域名或一个英文单词,访问一个字典的条目。映射(map)、哈希表(hashmap)、哈希(hash)、关联数组(associative array)是字典的其他名字。在一般情况下,我们希望能够在任意类型的信息之间映射。各种语言学对象以及它们的映射见表 5-3。

表 5-3 语言学对象从键到值的映射

语言学对象	映射来自	映射到
文档索引	词	页面列表(找到词的地方)
同义词	词义	同义词列表
词典	中心词	词条项(词性、意思定义、词源)
比较单词列表	注释术语	同源词(词列表,每种语言一个)
词形分析	表面形式	形态学分析(词素组件列表)

5.3.2 Python 字典

Python 提供了一个字典数据类型,可用来做任意类型之间的映射,它更像是一个传统的字典,以一种高效的方式来查找事物。定义 pos 为一个空字典{},然后给它添加 4 个项目,指定一些词的词性,使用数组方括号[]将条目添加到字典。

```
>>> pos={}
>>> pos
{}
```

```
>>> pos['colorless']='ADJ'
>>> pos
{'colorless': 'ADJ'}
>>> pos['ideas']='N'
>>> pos['sleep']='V'
>>> pos['furiously']='ADV'
>>> pos #注意顺序的变化
{'furiously': 'ADV', 'sleep': 'V', 'ideas': 'N', 'colorless': 'ADJ'}
>>> pos={'furiously':'ADV','sleep':'V','ideas':'N','colorless':'ADJ'}
>>> pos
{'furiously': 'ADV', 'sleep': 'V', 'ideas': 'N', 'colorless': 'ADJ'}
>>> pos['file']
Traceback (most recent call last):
  File "<stdin>", line 1, in <module>
KeyError: 'file'
>>> pos['file']='N'
```

示例中，colorless 的词性是形容词，或者更具体地说，在字典 pos 中，键'colorless'分配了值'ADJ'，当检查 pos 的值时，可以看到一个键值对的集合，一旦以这样的方式填充了字典，就可以使用键来检索值。

```
>>> pos['ideas'] #'N'
>>> pos['colorless'] #'ADJ'
```

当然，可能会无意中使用一个尚未分配值的键。

```
>>> pos['green']
Traceback (most recent call last):
  File "<stdin>", line 1, in <module>
KeyError: 'green'
```

这就提出了一个重要的问题。与列表和字符串中可以用 len()算出哪些整数是合法索引不同，我们如何算出一个字典的合法键？如果字典不是太大，可以简单地通过查看变量 pos 检查它的内容。

```
>>> list(pos)
['furiously', 'sleep', 'ideas', 'colorless']
>>> sorted(pos)
['colorless', 'furiously', 'ideas', 'sleep']
>>> [w for w in pos if w.endswith('s')]
['ideas', 'colorless']
```

当输入 list(pos)时，看到的结果可能会与这里显示的顺序不同。如果想看到有序的键，只需要对它们进行排序。与使用一个 for 循环遍历字典中的所有键一样，可以使用 for 循环输出字典的内容。

```
for word in sorted(pos):
    print word+":", pos[word]
colorless: ADJ
furiously: ADV
ideas: N
sleep: V
```

最后，字典的方法 keys()、values()和 items()允许访问作为单独的列表的键、值以及键-值对。甚至可以按它们的第一个元素排序元组（如果第一个元素相同，就使用它们的第二个元素）。

```
pos.keys()
pos.values()
pos.items()
for key, val in sorted(pos.items()):
    print key+":", val
colorless: ADJ
furiously: ADV
ideas: N
sleep: V
```

要确保在字典中查找某词时，一个键只得到一个值。现在假设用字典来存储可同时作为动词和名词的词 sleep。

```
>>> pos['sleep']='V'
>>> pos['sleep'] #'V'
>>> pos['sleep']='N'
>>> pos['sleep'] #'N'
```

一开始，pos['sleep']给的值是'V'。但是，它立即被一个新值'N'覆盖了。换句话说，字典中只能有'sleep'的一个条目。然而，有一个方法可以在该项目中存储多个值：使用一个列表值，如 pos['sleep']=['N', 'V']。

```
>>> pos
{'furiously': 'ADV', 'sleep': ['N', 'V'], 'ideas': 'N', 'colorless': 'ADJ'}
```

事实上，这就是 CMU 发音字典，它为一个词存储多个发音。

5.3.3 定义字典

可以使用键值对格式创建字典。

```
>>> pos={'colorless':'ADJ', 'ideas':'N', 'sleep':'V', 'furiously':'ADV'}
>>> pos
{'furiously': 'ADV', 'sleep': 'V', 'ideas': 'N', 'colorless': 'ADJ'}
>>> pos=dict(colorless='ADJ', ideas='N', sleep='V', furiously='ADV')
>>> pos
{'furiously': 'ADV', 'sleep': 'V', 'ideas': 'N', 'colorless': 'ADJ'}
```

注意，字典的键必须是不可改变的类型，如字符串和元组。如果尝试使用可变键定义字典，则会得到一个 TypeError。

```
>>> pos={['ideas', 'blogs', 'adventures']: 'N'} #字典长度可变
Traceback (most recent call last):
  File "<stdin>", line 1, in <module>
TypeError: unhashable type: 'list'
>>> pos={('ideas', 'blogs', 'adventures'): 'N'} #元组分量数、顺序都不可改变
>>> pos
{('ideas', 'blogs', 'adventures'): 'N'}
```

如果访问一个不在字典中的键，则会得到一个错误。然而，如果一个字典能为这个新键自动创建一个条目并给它一个默认值（如 0 或者一个空列表），将是非常有用的（类似于动态查找表）。自从 Python 2.5 以来，一种特殊的称为 defaultdict 的字典已经出现。考虑到有的读者使用 Python 2.4，NLTK 提供了 nltk.defaultdict。为了使用它，必须提供一个参数用来创建默认值，如 int、float、str、list、dict、tuple。

```
import nltk
frequency=nltk.defaultdict(int)
frequency['colorless']=4
frequency['ideas'] #0
pos=nltk.defaultdict(list)
pos['sleep']=['N', 'V']
pos['ideas'] #[]
```

这些默认值实际上是将其他对象转换为指定类型的函数，如 int("2")、list("2")。它们被调用时没有参数，int()、list()分别返回 0 和[]。

前面的例子指定字典项的默认值为一个特定的数据类型的默认值。然而，也可以指定其他的默认值，创建一个任一条目的默认值是'N'的字典，当访问一个不存在的条目时，它会自动添加到字典。

```
pos=nltk.defaultdict(lambda:'N')
pos['colorless']='ADJ'
pos['blog'] #'N'
pos.items() #[('blog', 'N'), ('colorless', 'ADJ')]
```

这个例子使用一个 lambda 表达式。这个 lambda 表达式没有指定参数，所以需要用不带参数的括号调用它。因此，下面的 f 和 g 的定义是等价的。

```
f=lambda:'N'
f() #'N'
def g():
    return 'N'


g() #'N'
```

让我们来看看默认字典如何被应用在较大规模的语言处理任务中。

创建一个默认字典，映射每个词为它们的替换词。最频繁的 n 个词将被映射到它们自己，其他的被映射到 UNK。

```
import nltk
from collections import Counter
alice=nltk.corpus.gutenberg.words('carroll-alice.txt')
vocab=nltk.FreqDist(alice)
v1000=list(vocab)[:1000] #vocab.keys()[:1000], vocab.most_common(1000)
mapping=nltk.defaultdict(lambda:'UNK') #
for v in v1000:
    mapping[v]=v


alice2=[mapping[v] for v in alice]
alice2[:100]
['[', 'Alice', "'", 's', 'Adventures', 'in', 'Wonderland', 'by',
'UNK', 'UNK', 'UNK', 'UNK', 'CHAPTER', 'I', '.', 'Down', 'the', 'Rabbit',
'-', 'UNK', 'Alice', 'was', 'beginning', 'to', 'get', 'very', 'tired',
'of', 'sitting', 'by', 'her', 'sister', 'on', 'the', 'bank', ',', 'and',
'of', 'having', 'nothing', 'to', 'do', ':', 'once', 'or', 'twice', 'she',
'had', 'UNK', 'into', 'the', 'bo', 'her', 'sister', 'was', 'UNK', ',
', 'but', 'it', 'had', 'no', 'pictures', 'or', 'UNK', 'in', 'it', ',
', "'", 'and', 'what', 'is', 'the', 'use', 'of', 'a', 'bo', '', "'",
```

```
'thought', 'Alice', "'", 'without', 'pictures', 'or', 'conversation',
"?'", 'So', 'she', 'was', 'UNK', 'in', 'her', 'own', 'mind', '(', 'as',
'well', 'as', 'she', 'could', ',', ']
    len(set(alice2))   #1001
```

5.3.4 递增地更新字典

可以使用字典计数词出现的次数。首先初始化一个空的 defaultdict，然后处理文本中每个词性标记，如果标记以前没有见过，就默认计数为零，每次遇到一个标记，就使用+=运算符递增它的计数。

递增地更新字典，按值排序。

```
from nltk import defaultdict
counts=defaultdict(int)
from nltk.corpus import brown
for word, tag in brown.tagged_words(categories='news', simplify_tags=True):
    counts[tag]+=1

counts['N']
list(counts)
from operator import itemgetter
#sorted(counts.items(), key=itemgetter(1), reverse=True)
def cmp(x, y):
    return y[1]-x[1]

sorted(counts.items(), cmp=cmp)
[t for t, c in sorted(counts.items(), key=itemgetter(1), reverse=True)]
```

按频率递减顺序显示词汇。sorted()的第一个参数是待排序列，它是由一个 POS 标记和一个频率组成的元组的列表，第二个参数使用函数 itemgetter()指定排序键，在一般情况下，itemgetter(n)返回一个函数，这个函数可以在一些其他序列对象上调用获得这个序列的第 n 个元素。

```
pair=('NP', 8336)
pair[1] #8336
itemgetter(1)(pair) #8336
```

sorted()的最后一个参数指定项目是否应按相反的顺序返回，即频率值递减。

初始化一个 defaultdict，然后使用 for 循环来更新其值。下面是这种模式的另一个实例，按它们最后两个字母索引词汇。

```
last_letters=nltk.defaultdict(list)
words=nltk.corpus.words.words('en')
for word in words:
    key=word[-2:]
    last_letters[key].append(word)

last_letters['ly'][:9]
```

下面的例子使用相同的模式创建一个颠倒顺序的词字典。可以试验第 3 行，以便能弄清楚为什么这个程序能运行。

```
anagrams=nltk.defaultdict(list)
for word in words:
    key=''.join(sorted(word))
    anagrams[key].append(word)

anagrams['aeilnrt']
```

NLTK 以 nltk.Index()的形式提供一个更方便创建 defaultdict(list)的方式。

```
import nltk
from nltk.corpus import words
words=words.words('en')
anagrams=nltk.Index((''.join(sorted(w)), w) for w in words)
anagrams['aeilnrt']
```

nltk.Index 是一个支持初始化的 defaultdict(list)，nltk.FreqDist 本质上是一个支持初始化的 defaultdict(int)，附带排序和绘图方法。

给定一组字符串，返回所有满足 Anagrams 的字符串。Anagrams 是指由颠倒字母顺序组成的单词，如 "dormitory" 颠倒字母顺序会变成 "dirty room"，"tea" 会变成 "eat"。

5.3.5　复杂的键和值

可以使用具有复杂的键和值的默认字典，下面研究一个词可能的标记范围。给定词本身和它前面一个词的标记，将看到这些信息如何被一个 POS 标注器使用。

```
>>> from nltk import defaultdict, bigrams
>>> from nltk.corpus import brown
>>> pos=defaultdict(lambda:defaultdict(int))
>>>     brown_news_tagged=brown.tagged_words(categories='news',
tagset='universal')
>>> for (w1, t1), (w2, t2) in bigrams(brown_news_tagged):
...     pos[(t1, w2)][t2]+=1
```

```
...
>>> pos[('DET', 'right')]
defaultdict(<type 'int'>, {u'ADJ': 11, u'NOUN': 5})
```

这个例子使用一个字典,它的条目默认值也是一个字典,其默认值是 int(),即 0。注意如何遍历已标注语料库的双连词,每次遍历处理一个词标记对,每次通过循环时,更新字典 pos 中的条目(t1, w2)、一个标记和它后面的词。当在 pos 中查找一个项目时,必须指定一个复合键,然后得到一个字典对象。一个 POS 标注器可以使用这些信息来决定词 right。前面是一个限定词时,应标注为 ADJ。

5.3.6 颠倒字典

字典支持高效查找,想要获得任意键的值,如果 d 是一个字典,k 是一个键,输入 d[k],就能立即获得值。

```
>>> from nltk import defaultdict
>>> from nltk.corpus import gutenberg
>>> counts=defaultdict(int)
>>> for w in gutenberg.words('milton-paradise.txt'):
...     counts[w]+=1
...
>>> [k for k, v in counts.items() if v==32]
[u'brought', u'Him', u'virtue', u'Against', u'There', u'thine',
u'King', u'mortal', u'every', u'been']
>>> counts.items()
[(u'voluble', 1), (u'Omnipotent', 11), (u'Hasting', 1), (u'foul', 26),

from nltk import defaultdict, word_tenize
counts=defaultdict(int)
text='This is a bo . That is a desk.'
text=word_tenize(text)
for word in text:
    counts[word]+=1

[key for key, value in counts.items() if value==2]
```

如果希望经常做这样的一种"反向查找",那么,建立一个映射值到键的字典是很有用的。在没有两个键具有相同的值情况下,这是一件容易的事。只要得到字典中的所有键-值对,并创建一个新的值-键对字典。下面演示用键-值对初始化字典 pos 的另一种方式。

```
pos={'colorless':'ADJ', 'ideas':'N', 'sleep':'V', 'furiously':'ADV'}
pos2=dict((value, key) for key, value in pos.items())
pos2['N'] #'ideas'
```

使用字典的 update()方法再加入一些词到 pos 中，创建多个键具有相同的值。这样一来，刚才看到的反向查找技术将不再起作用。作为替代，不得不使用 append()积累词和每个词性。

```
pos={'colorless':'ADJ', 'ideas':'N', 'sleep':'V', 'furiously':'ADV'}
pos2=dict((value, key) for key, value in pos.items())
pos2['N'] #'ideas'
pos.update({'cats':'N', 'scratch':'V', 'peacefully':'ADV', 'old':'ADJ'})
pos2=nltk.defaultdict(list) # defaultdict(<type 'list'>, {})
for key, value in pos.items():
    pos2[value].append(key)

pos2['ADV'] #['peacefully', 'furiously']
```

现在，已经颠倒了字典 pos，可以查任意词性，找到所有具有此词性的词。可以使用 NLTK 中的索引。

```
pos={'colorless':'ADJ', 'ideas':'N', 'sleep':'V', 'furiously':'ADV'}
pos.update({'cats':'N', 'scratch':'V', 'peacefully':'ADV', 'old':'ADJ'})
pos2=nltk.Index((value, key) for key, value in pos.items()) #Δ
pos2['ADV']
```

表 5-4 给出了 Python 字典方法总结。

表 5-4 Python 字典常用方法与字典相关习惯用法

示例	说明
d={}	创建一个空的字典，并分配给 d
d[key]=value	分配一个值给一个给定的字典键
d.keys()	字典的键的列表
list(d)	字典的键的列表
sorted(d)	字典的键的排序
key in d	测试一个特定的键是否在字典中
for key in d	遍历字典的键
d.values()	字典中的值的列表
dict([(k1, v1), (k2, v2), ...])	从一个键-值对列表创建一个字典
d1.update(d2)	添加 d2 中所有项目到 d1
defaultdict(int)	一个默认值为 0 的字典

5.4 自动标注

下面将探讨以不同的方式来给文本自动添加词性标记。一个词的标记依赖于这个词和它在句子中的上下文，出于这个原因，我们将处理句子层次而不是词汇层次的数据。

5.4.1 默认标注器

最简单的标注器是为每个标识符分配同样的标记，这似乎是一个相当普通的一步，但它建立了标注器性能的一个重要的下界。为了得到最好的效果，我们用最有可能的标记标注每个词。

```
from nltk.corpus import brown
brown_tagged_sents=brown.tagged_sents(categories='news')
brown_sents=brown.sents(categories='news')
from nltk import FreqDist
tags=[tag for (_, tag) in brown.tagged_words(categories='news')]
FreqDist(tags).max()
```

创建一个将所有词都标注成 NN 的标注器。

```
from nltk import word_tokenize, DefaultTagger, pos_tag
raw='I do not like green eggs and ham , I do not like them Sam I am ! '
tokens=word_tokenize(raw)
default_tagger=DefaultTagger('NN')
default_tagger.tag(tokens)
tagged_words=pos_tag(tokens)
default_tagger.evaluate([tagged_words])
default_tagger.evaluate(brown_tagged_sents)
```

```
Python 2.7.8 |Anaconda 2.1.0 (64-bit)| (default, Jul  2 2014,
15:12:11) [MSC v.1500 64 bit (AMD64)] on win32
Type "help", "copyright", "credits" or "license" for more information.
Anaconda is brought to you by Continuum Analytics.
Please check out: http://continuum.io/thanks and https://binstar.org
>>> from nltk.corpus import brown
>>> brown_tagged_sents=brown.tagged_sents(categories='news')
>>> brown_sents=brown.sents(categories='news')
>>> from nltk import FreqDist
```

```
>>> tags=[tag for (_, tag) in brown.tagged_words(categories='news')]
>>> FreqDist(tags).max()
u'NN'
>>> from nltk import word_tokenize, DefaultTagger, pos_tag
>>> raw='I do not like green eggs and ham , I do not like them Sam I am ! '
>>> tokens=word_tokenize(raw)
>>> default_tagger=DefaultTagger('NN')
>>> default_tagger.tag(tokens)
[('I', 'NN'), ('do', 'NN'), ('not', 'NN'), ('like', 'NN'), ('green', 'NN'), ('eggs', 'NN'), ('and', 'NN'), ('ham', 'NN'), (',', 'NN'), ('I', 'NN'), ('do', 'NN'), ('not', 'NN'), ('like', 'NN'), ('them', 'NN'), ('Sam', 'NN'), ('I', 'NN'), ('am', 'NN'), ('!', 'NN')]
>>> tagged_words=pos_tag(tokens)
>>> default_tagger.evaluate([tagged_words])
0.05555555555555555
>>> default_tagger.evaluate(brown_tagged_sents)
0.13089484257215028
>>>
```

在一个典型的语料库中，它只标注正确了 1/8 的标识符，默认的标注器给每一个单独的词分配标记。这意味着默认标注器可以帮助提高语言处理系统的稳定性。

```
from nltk import word_tokenize, DefaultTagger, pos_tag
tagged_sents=[[('This', 'AT'), ('is', 'NP-TL'), ('a', 'NN-TL'), ('bo', 'NN'), ('.', '.')], [('That', 'AT'), ('is', 'NP-TL'), ('a', 'NN-TL'), ('desk', 'NN'), ('.', '.')]]
sents=[['This', 'is', 'a', 'book', '.'], ['That', 'is', 'a', 'desk', '.']]
raw='I do not like green eggs and ham, I do not like them Sam I am!'
tokens=word_tokenize(raw)
default_tagger=DefaultTagger('NN')
default_tagger.tag(tokens)
default_tagger.tag(sents[0])
default_tagger.evaluate(tagged_sents)

>>> from nltk import word_tokenize, DefaultTagger, pos_tag
>>> tagged_sents=[[('This', 'AT'), ('is', 'NP-TL'), ('a', 'NN-TL'), ('bo', 'NN'), ('.', '.')], [('That', 'AT'), ('is', 'NP-TL'), ('a', 'NN-TL'), ('desk', 'NN'), ('.', '.')]]
>>> sents=[['This', 'is', 'a', 'book', '.'], ['That', 'is', 'a', 'desk', '.']]
```

```
>>> raw='I do not like green eggs and ham, I do not like them Sam I am!'
>>> tokens=word_tokenize(raw)
>>> default_tagger=DefaultTagger('NN')
>>> default_tagger.tag(tokens)
[('I', 'NN'), ('do', 'NN'), ('not', 'NN'), ('like', 'NN'), ('green',
'NN'), ('eggs', 'NN'), ('and', 'NN'), ('ham', 'NN'), (',', 'NN'
), ('I', 'NN'), ('do', 'NN'), ('not', 'NN'), ('like', 'NN'), ('them',
'NN'), ('Sam', 'NN'), ('I', 'NN'), ('am', 'NN'), ('!', 'NN')]
>>> default_tagger.tag(sents[0])
[('This', 'NN'), ('is', 'NN'), ('a', 'NN'), ('book', 'NN'), ('.', 'NN')]
>>> default_tagger.evaluate(tagged_sents)
0.2
```

5.4.2 正则表达式标注器

正则表达式标注器基于匹配模式分配标记给标识符，例如，可能会猜测任一以 ed 结尾的词是动词的过去分词（VBD），任一以's 结尾的词是名词的所有格（NN$）。可以用一个正则表达式的列表来表示这些。

```
>>> from nltk.corpus import brown
>>> brown_tagged_sents=brown.tagged_sents(categories='news')
>>> brown_sents=brown.sents(categories='news')
>>> from nltk import RegexpTagger #RegexpParser, re, compile
>>> patterns=[(r'.*ing$', 'VBG'), #gerunds 动名词, {}
... (r'.*ed$', 'VBD'), #simple past
... (r'.*es$', 'VBZ'), #3rd singular present
... (r'.*ould$', 'MD'), #modals
... (r'.*\'s$', 'NN$'), #possessive nouns
... (r'.*s$', 'NNS'), #plural nouns
... (r'^-?[0-9]+(.[0-9]+)?$', 'CD'), #cardinal numbers
... (r'.*', 'NN') #nouns(default)
... ]
>>> regexp_tagger=RegexpTagger(patterns)
>>> regexp_tagger.tag(brown_sents[3])
[(u'``', 'NN'), (u'Only', 'NN'), (u'a', 'NN'), (u'relative', 'NN'),
(u'handful', 'NN'), (u'of', 'NN'), (u'such', 'NN'), (u'reports', 'NNS'),
(u'was', 'NNS'), (u'received', 'VBD'), (u"'", 'NN'), (u',', 'NN'),
(u'the', 'NN'), (u'jury', 'NN'), (u'said', 'NN'), (u',', 'NN'), (u'``',
```

```
'NN'), (u'considering', 'VBG'), (u'the', 'NN'), (u'widespread', 'NN'),
(u'interest', 'NN'), (u'in', 'NN'), (u'the', 'NN'), (u'election', 'NN'),
(u',', 'NN'), (u'the', 'NN'), (u'number', 'NN'), (u'of', 'NN'),
(u'voters', 'NNS'), (u'and', 'NN'), (u'the', 'NN'), (u'size', 'NN'),
(u'of', 'NN'), (u'this', 'NNS'), (u'city', 'NN'), (u"'", 'NN'), (u'.',
'NN')]
    >>> regexp_tagger.evaluate(brown_tagged_sents)
    0.20326391789486245
    >>> brown_tagged_sents[3]
    [(u'``', u'``'), (u'Only', u'RB'), (u'a', u'AT'), (u'relative', u'JJ'),
(u'handful', u'NN'), (u'of', u'IN'), (u'such', u'JJ'), (u'reports',
u'NNS'), (u'was', u'BEDZ'), (u'received', u'VBN'), (u"'", u"'"), (u',
', u','), (u'the', u'AT'), (u'jury', u'NN'), (u'said', u'VBD'), (u',
', u','), (u'``', u'``'), (u'considering', u'IN'), (u'the', u'AT'),
(u'widespread', u'JJ'), (u'interest', u'NN'), (u'in', u'IN'), (u'the',
u'AT'), (u'election', u'NN'), (u',', u','), (u'the', u'AT'), (u'number',
u'NN'), (u'of', u'IN'), (u'voters', u'NNS'), (u'and', u'CC'), (u'the',
u'AT'), (u'size', u'NN'), (u'of', u'IN'), (u'this', u'DT'), (u'city',
u'NN'), (u"'", u"'"), (u'.', u'.')]
```

注意：这些是顺序处理的。第一个匹配上的会被使用，可以建立一个标注器，并用它来标记一个句子。做完这一步会有约 1/5 是正确的。最终的正则表达式.*是一个全面捕捉的，标注所有词为名词。除了作为正则表达式标注器的一部分被重新指定，这与默认标注器是等效的，只是效率较低。

5.4.3 查询标注器

很多高频词没有 NN 标记。找出 100 个最频繁的词，存储它们最有可能的标记，然后可以使用这个信息作为查询标注器的模型。

```
Python 2.7.11 |Anaconda 2.5.0 (64-bit)| (default, Jan 29 2016,
14:26:21) [MSC v.1500 64 bit (AMD64)] on win32
Type "help", "copyright", "credits" or "license" for more information.
Anaconda is brought to you by Continuum Analytics.
Please check out: http://continuum.io/thanks and https://anaconda.org
    >>> from nltk import FreqDist, ConditionalFreqDist, UnigramTagger
    >>> from nltk.corpus import brown
    >>> brown_tagged_sents=brown.tagged_sents(categories='news')
    >>> fd=FreqDist(brown.words(categories='news'))
```

```
>>> cfd=ConditionalFreqDist(brown.tagged_words(categories='news'))
>>> most_freq_words=[w for w, _ in fd.most_common(100)]
>>> likely_tags=dict((word, cfd[word].max()) for word in most_freq_words)
>>> baseline_tagger=UnigramTagger(model=likely_tags)
>>> baseline_tagger=UnigramTagger([likely_tags.items()])
>>> baseline_tagger.evaluate(brown_tagged_sents)
0.45578495136941344
>>>
>>>
```

仅仅知道 100 个最频繁词的标记就使我们能正确标注很大一部分标识符。让我们来看看它在一些未标注的输入文本上做得如何。

```
sent=brown.sents(categories='news')[3]
baseline_tagger.tag(sent)
```

许多词都被分配了一个 None 标签，因为它们不在 100 个最频繁的词之中。在这些情况下，我们想分配默认标记 NN。换句话说，要先使用查找表，如果它不能指定一个标记就使用默认标注器，这个过程叫作回退。通过指定一个标注器作为另一个标注器的参数，如下所示。现在查找标注器将只存储名词以外的词的词-标记对，只要它不能给一个词分配标记，它将会调用默认标注器。

```
baseline_tagger=nltk.UnigramTagger(model=likely_tags,
backoff=nltk.DefaultTagger('NN'))
```

查找标注器的性能，使用不同大小的模型。

```
import nltk
from nltk.corpus import brown
def performance(cfd, wordlist):
    lt=dict((word, cfd[word].max()) for word in wordlist)
    #baseline_tagger=nltk.UnigramTagger(model=lt,  backoff=nltk.DefaultTagger('NN'))
    baseline_tagger=nltk.UnigramTagger(model=lt)
    return baseline_tagger.evaluate(brown.tagged_sents(categories='news'))

def display():
    import pylab
    words_by_freq=list(nltk.FreqDist(brown.words(categories='news')))
    cfd=nltk.ConditionalFreqDist(brown.tagged_words(categories='news'))
    sizes=2**pylab.arange(15) #2[0, 1, 2, 3, 4, 5, 6, 7, 8, 9, 10, 11, 12, 13, 14]=[1, 2, 4, 8, 16, 32, 64, 128, 256, 512, 1024, ]
```

```
    perfs=[performance(cfd, words_by_freq[:size]) for size in sizes]
    pylab.plot(sizes, perfs, '-bo')
    pylab.title('Loup Tagger Performance with Varying Model Size')
    pylab.xlabel('Model Size')
    pylab.ylabel('Performance')
    pylab.show()

display()
```

观察图 5-3，随着模型规模的增长，最初的性能增加迅速，最终会达到一个稳定水平，这时模型的规模大量增加，但性能的提高却很小。

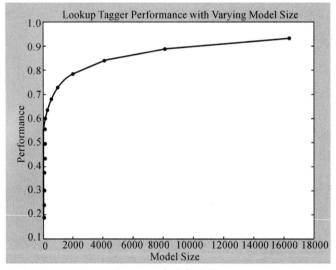

图 5-3　查找标注器

5.5　N-gram 标注

5.5.1　一元标注(Unigram Tagging)

一元标注器基于一个简单的统计算法，对每个标识符分配最有可能的标记，例如，它将分配标记 JJ 给词 frequent，因为 frequent 用作一个形容词比用作一个动词更常见。一元标注器的行为就像一个查找标注器，有一个更方便建立它的技术，称为训练。在下面的例子中，训练一元标注器，用它来标注一个句子，然后进行评估。

```
from nltk import UnigramTagger
from nltk.corpus import brown
brown_tagged_sents=brown.tagged_sents(categories='news')
```

```
brown_sents=brown.sents(categories='news')
unigram_tagger=UnigramTagger(brown_tagged_sents) #训练
brown_tagged_sents2007=unigram_tagger.tag(brown_sents[2007])
unigram_tagger.evaluate([brown_tagged_sents2007])
unigram_tagger.evaluate(brown_tagged_sents)
>>> from nltk import UnigramTagger
>>> from nltk.corpus import brown
>>> brown_tagged_sents=brown.tagged_sents(categories='news')
>>> brown_sents=brown.sents(categories='news')
>>> unigram_tagger=UnigramTagger(brown_tagged_sents) #训练
>>> brown_tagged_sents2007=unigram_tagger.tag(brown_sents[2007])
>>> unigram_tagger.evaluate([brown_tagged_sents2007])
1.0
>>> unigram_tagger.evaluate(brown_tagged_sents)
0.9349006503968017
>>> brown_tagged_sents2007
[('Various', 'JJ'), ('of', 'IN'), ('the', 'AT'), ('apartments',
'NNS'), ('are', 'BER'), ('of', 'IN'), ('the', 'AT'), ('terrace', 'NN'),
('type', 'NN'), (',', ','), ('being', 'BEG'), ('on', 'IN'), ('the',
'AT'), ('ground', 'NN'), ('floor', 'NN'), ('so', 'QL'), ('that', 'CS'),
('entrance', 'NN'), ('is', 'BEZ'), ('direct', 'JJ'), ('.', '.')]
>>> brown_tagged_sents[2007]
[('Various', 'JJ'), ('of', 'IN'), ('the', 'AT'), ('apartments',
'NNS'), ('are', 'BER'), ('of', 'IN'), ('the', 'AT'), ('terrace', 'NN'),
('type', 'NN'), (',', ','), ('being', 'BEG'), ('on', 'IN'), ('the',
'AT'), ('ground', 'NN'), ('floor', 'NN'), ('so', 'CS'), ('that', 'CS'),
('entrance', 'NN'), ('is', 'BEZ'), ('direct', 'JJ'), ('.', '.')]
```

训练 UnigramTagger，通过在初始化标注器时指定已标注的句子集数据作为参数，训练过程中涉及检查每个词的标记，将所有词最可能的标记存储在一个字典里面，这个字典存储在标注器内部。

5.5.2 分离训练和测试数据

现在，我们正在一些数据上训练一个标注器，必须小心，不要在相同的数据上进行测试。分割数据，90%为训练数据，其余10%为测试数据。

```
>>> from nltk.corpus import brown
>>> from nltk import UnigramTagger
```

```
>>> brown_tagged_sents=brown.tagged_sents()
>>> size=int(len(brown_tagged_sents)*0.005)
>>> brown_tagged_sents=brown_tagged_sents[:size]
>>> size=int(len(brown_tagged_sents)*0.9)
>>> size
257
>>> train_sents=brown_tagged_sents[:size]
>>> test_sents=brown_tagged_sents[size:]
>>> unigram_tagger=UnigramTagger(train_sents)
>>> unigram_tagger.evaluate(train_sents)
0.9546776124508295
>>> unigram_tagger.evaluate(test_sents)
0.7476979742173112
```

虽然得分更糟糕了，但是我们现在对这种标注器是无用的有了一个更好的了解。

5.5.3　N-gram 标注

一个 n-gram 标注器是一个 unigram 标注器的一般化，它的上下文是当前词和它前面 n-1 个标识符的词性标记。在 n-gram 标注器的例子中，如果让 n=3，也就是说，我们考虑当前词的前 2 个词的标记。一个 n-gram 标注器挑选在给定的上下文中最有可能的标记，1-gram 标注器是一元标注器，即用于标注一个标识符的上下文只是标识符本身，2-gram 标注器也称为二元标注器（bigram taggers），3-gram 标注器也称为三元标注器（trigram taggers）。NgramTagger 类使用一个已标注的训练语料库来确定对每个上下文哪个词性标记最有可能。在这里，可以看到一个 n-gram 标注器的特殊情况，即一个 bigram 标注器。训练它，用它来标注未标注的句子。

```
Python 2.7.11 |Anaconda 2.5.0 (64-bit)| (default, Jan 29 2016, 14:26:21) [MSC v.1500 64 bit (AMD64)] on win32
>>> from nltk import BigramTagger
>>> from nltk.corpus import brown
>>> brown_tagged_sents=brown.tagged_sents(categories='news')
>>> brown_sents=brown.sents(categories='news')
>>> size=int(len(brown_tagged_sents)*0.9)
>>> train_sents=brown_tagged_sents[:size]
>>> test_sents=brown_tagged_sents[size:]
>>> bigram_tagger=BigramTagger(train_sents)
>>> bigram_tagger.tag(brown_sents[2007])
[(u'Various', u'JJ'), (u'of', u'IN'), (u'the', u'AT'), (u'apartments',
```

```
u'NNS'), (u'are', u'BER'), (u'of', u'IN'), (u'the', u'AT'), (u'terrace',
u'NN'), (u'type', u'NN'), (u',', u','), (u'being', u'BEG'), (u'on',
u'IN'), (u'the', u'AT'), (u'ground', u'NN'), (u'floor', u'NN'), (u'so',
u'CS'), (u'that', u'CS'), (u'entrance', u'NN'), (u'is', u'BEZ'),
(u'direct', u'JJ'), (u'.', u'.')]
>>> unseen_sent=brown_sents[4203]
>>> bigram_tagger.tag(unseen_sent)
[(u'The', u'AT'), (u'population', u'NN'), (u'of', u'IN'), (u'the',
u'AT'), (u'Congo', u'NP'), (u'is', u'BEZ'), (u'13.5', None), (u'million',
None), (u',', None), (u'divided', None), (u'into', None), (u'at', None),
(u'least', None), (u'seven', None), (u'major', None), (u'``', None),
(u'culture', None), (u'clusters', None), (u"''", None), (u'and', None),
(u'innumerable', None), (u'tribes', None), (u'speaking', None), (u'400',
None), (u'separate', None), (u'dialects', None), (u'.', None)]
>>> bigram_tagger.evaluate(test_sents)
0.10276088906608193
>>> train_sents
[[(u'The', u'AT'), (u'Fulton', u'NP-TL'), (u'County', u'NN-TL'),
(u'Grand', u'JJ-TL'), (u'Jury', u'NN-TL'), (u'said', u'VBD'), (u'Friday',
u'NR'), (u'an', u'AT'), (u'investigation', u'NN'), (u'of', u'IN'),
(u"Atlanta's", u'NP$'), (u'recent', u'JJ'), (u'primary', u'NN'),
(u'election', u'NN'), (u'produced', u'VBD'), (u'``', u'``'), (u'no',
u'AT'), (u'evidence', u'NN'), (u"''", u"''"), (u'that', u'CS'), (u'any',
u'DTI'), (u'irregularities', u'NNS'), (u'took', u'VBD'), (u'place',
u'NN'), (u'.', u'.')], [(u'The', u'AT'), (u'jury', u'NN'), (u'further',
u'RBR'), (u'said', u'VBD'), (u'in', u'IN'), (u'term-end', u'NN'),
(u'presentments', u'NNS'), (u'that', u'CS'), (u'the', u'AT'), (u'City',
u'NN-TL'), (u'Executive', u'JJ-TL'), (u'Committee', u'NN-TL'), (u',',
u','), (u'which', u'WDT'), (u'had', u'HVD'), (u'over-all', u'JJ'),
(u'charge', u'NN'), (u'of', u'IN'), (u'the', u'AT'), (u'election',
u'NN'), (u',', u','), (u'``', u'``'), (u'deserves', u'VBZ'), (u'the',
u'AT'), (u'praise', u'NN'), (u'and', u'CC'), (u'thanks', u'NNS'),
(u'of', u'IN'), (u'the', u'AT'), (u'City', u'NN-TL'), (u'of', u'IN-TL'),
(u'Atlanta', u'NP-TL'), (u"''", u"''"), (u'for', u'IN'), (u'the',
u'AT'), (u'manner', u'NN'), (u'in', u'IN'), (u'which', u'WDT'), (u'the',
u'AT'), (u'election', u'NN'), (u'was', u'BEDZ'), (u'conducted',
u'VBN'), (u'.', u'.')], ...]
>>> type(train_sents)
<class 'nltk.util.LazySubsequence'>
```

```
>>> brown_sents[2007]
[u'Various', u'of', u'the', u'apartments', u'are', u'of', u'the',
u'terrace', u'type', u',', u'being', u'on', u'the', u'ground',
u'floor', u'so', u'that', u'entrance', u'is', u'direct', u'.']
```

注意，bigram 标注器能够标注训练中它看到过的句子中的所有词，但对一个没见过的句子表现很差。只要遇到一个新词，就无法给它分配标记。它不能标注下面的词，即使是在训练过程中看到过的。因此，标注器标注句子的其余部分也失败了。它的整体准确度得分非常低。

```
bigram_tagger.evaluate(test_sents)
0.10216286255357321
```

当 n 越大，上下文的特异性就会增加，要标注的数据中包含训练数据中不存在的上下文的概率也增大。这被称为数据稀疏问题，在 NLP 中是相当普遍的。因此，我们的研究结果的精度和覆盖范围需要有一个权衡（这与信息检索中的精度/召回权衡有关）。

注意：N-gram 标注器不应考虑跨越句子边界的上下文。因此，NLTK 的标注器被设计用于句子列表，一个句子是一个词列表。在一个句子的开始，tn-1 和前面的标记被设置为 None。

5.5.4 组合标注器

解决精度和覆盖范围的一个办法是尽可能地使用更精确的算法，例如，可以按如下方式组合 bigram 标注器、unigram 标注器和一个默认标注器。

（1）尝试使用 bigram 标注器标注标识符。
（2）如果 bigram 标注器无法找到一个标记，尝试 unigram 标注器。
（3）如果 unigram 标注器也无法找到一个标记，使用默认标注器。

大多数 NLTK 标注器允许指定一个回退标注器，回退标注器自身可能也有一个回退标注器。

```
>>> from nltk.corpus import brown
>>> from nltk import DefaultTagger, UnigramTagger, BigramTagger
>>> brown_tagged_sents=brown.tagged_sents(categories='news')
>>> brown_sents=brown.sents(categories='news')
>>> size=int(len(brown_tagged_sents)*0.9)
>>> train_sents=brown_tagged_sents[:size]
>>> test_sents=brown_tagged_sents[size:]
>>> t0=DefaultTagger('NN')
>>> t1=UnigramTagger(train_sents, backoff=t0)
>>> t2=BigramTagger(train_sents, backoff=t1)
>>> t2.evaluate(test_sents)
0.844911791089405
```

注意：在标注器初始化时指定回退标注器，从而使训练能利用回退标注器。通过定义一个名为 t3 的 TrigramTagger，扩展前面的例子，它指定 t2 为回退标注器。

```
>>> from nltk.corpus import brown
>>> from nltk import DefaultTagger, UnigramTagger, BigramTagger, TrigramTagger
>>> brown_tagged_sents=brown.tagged_sents(categories='news')
>>> brown_sents=brown.sents(categories='news')
>>> size=int(len(brown_tagged_sents)*0.9)
>>> train_sents=brown_tagged_sents[:size]
>>> test_sents=brown_tagged_sents[size:]
>>> t0=DefaultTagger('NN')
>>> t1=UnigramTagger(train_sents, backoff=t0)
>>> t2=BigramTagger(train_sents, backoff=t1)
>>> t2.evaluate(test_sents)
0.844911791089405
>>> t3=TrigramTagger(train_sents, backoff=t2)
>>> t3.evaluate(test_sents)
0.8424200139539519
```

5.5.5 存储标注器

在大语料库上训练一个标注器可能需要大量的时间，没有必要在每次需要的时候重新训练一个标注器，可以将一个训练好的标注器保存到一个文件以后重复使用。保存标注器 t2 到文件 t2.pkl 中。

```
>>> from nltk import DefaultTagger, UnigramTagger, BigramTagger
>>> from nltk.corpus import brown
>>> brown_tagged_sents=brown.tagged_sents(categories='news')
>>> brown_sents=brown.sents(categories='news')
>>> size=int(len(brown_tagged_sents)*0.9)
>>> train_sents=brown_tagged_sents[:size]
>>> test_sents=brown_tagged_sents[size:]
>>> t0=DefaultTagger('NN')
>>> t1=UnigramTagger(train_sents, backoff=t0)
>>> t2=BigramTagger(train_sents, backoff=t1)
>>> t2.evaluate(test_sents)
0.844911791089405
>>> from cPickle import dump
```

```
>>> output=open('t2.pkl', 'wb')
>>> dump(t2, output, -1)
>>> output.close()
```

可以在一个单独的 Python 进程中载入保存的标注器。

```
>>> from cPickle import load
>>> input=open('t2.pkl', 'rb')
>>> tagger=load(input)
>>> input.close()
```

检查它是否可以用来标注。

```
>>> from cPickle import load
>>> input=open('t2.pkl', 'rb')
>>> tagger=load(input)
>>> input.close()
>>> text="""The board's action shows what free enterprise is up against in our complex maze of regulatory laws."""
>>> tokens=text.split()
>>> tagger.tag(tokens)
[('The', u'AT'), ("board's", u'NN$'), ('action', 'NN'), ('shows', u'NNS'), ('what', u'WDT'), ('free', u'JJ'), ('enterprise', 'NN'), ('is', u'BEZ'), ('up', u'RP'), ('against', u'IN'), ('in', u'IN'), ('our', u'PP$'), ('complex', u'JJ'), ('maze', 'NN'), ('of', u'IN'), ('regulatory', 'NN'), ('laws.', u'NN')]
```

5.5.6 性能限制

一个 n-gram 标注器的性能下限是什么？一个 trigram 标注器会遇到多少词性歧义的情况？可以根据经验回答这个问题。

```
from __future__ import division
from nltk import ConditionalFreqDist, trigrams
from nltk.corpus import brown
brown_tagged_sents=brown.tagged_sents(categories='news')
cfd=ConditionalFreqDist(((x[1], y[1], z[0]), z[1]) for sent in brown_tagged_sents for x, y, z in trigrams(sent)) #三连词
#TrigramTagger(train_sents)
ambiguous_contexts=[c for c in cfd.conditions() if len(cfd[c])>1]
sum(cfd[c].N() for c in ambiguous_contexts)/cfd.N() #0.049297702068029296
```

```
x=[('a', 'x'), ('b', 'y'), ('a', 'y'), ('a', 'x')]
y=ConditionalFreqDist(x)
y.N() #4
y['a'].items() #[('x', 2), ('y', 1)]
y['a'].N() #=3, [('a', 'x'), ('a', 'x'), ('a', 'y')]
len(y['a']) #=2, [('x', 2), ('y', 1)]
y['b'].items() #[('y', 1)]
```

因此，5%的 trigrams 是有歧义的。

测试标注器性能的另一种方法是研究它的错误。有些标记可能会比别的更难分配，可能需要专门对这些数据进行预处理或后处理。一种方便查看标注错误的方式是混淆矩阵。它用图表表示期望的标记（黄金标准）与实际由标注器产生的标记。

```
from __future__ import division
test_tags=[tag for sent in brown.sents(categories='editorial')
for (word, tag) in t2.tag(sent)]
gold_tags=[tag for (word, tag) in brown.tagged_words(categories='editorial')]
print nltk.ConfusionMatrix(gold_tags, test_tags)
```

5.5.7 跨句子边界标注

n-gram 标注器使用最近的标记作为当前词选择标记的指导。当标记一个句子的第一个词时，trigram 标注器将使用前面两个标识符的词性标记，通常会是前面句子的最后一个词和句子结尾的标点符号。然而，前一句结尾词的类别与下一句的开头通常没有关系，为了应对这种情况，可以使用已标注句子的列表来训练、运行和评估标注器。

句子层面的 n-gram 标注。

```
>>> from nltk import DefaultTagger, UnigramTagger, BigramTagger
>>> from nltk.corpus import brown
>>> brown_tagged_sents=brown.tagged_sents(categories='news')
>>> brown_sents=brown.sents(categories='news')
>>> size=int(len(brown_tagged_sents)*0.9)
>>> train_sents=brown_tagged_sents[:size]
>>> test_sents=brown_tagged_sents[size:]
>>> t0=DefaultTagger('NN')
>>> t1=UnigramTagger(train_sents, backoff=t0)
>>> t2=BigramTagger(train_sents, backoff=t1)
>>> t2.evaluate(test_sents)
0.844911791089405
```

5.6 基于转换的标注 Brill

Brill 标注是一种基于转换的学习，以它的发明者命名，猜每个词的标记，然后返回和修复错误。在这种方式中，Brill 标注器陆续将一个不良标注的文本转换成一个更好的，与 n-gram 标注一样，这是有监督的学习方法，因为需要已标注的训练数据来评估标注器的猜测是否是一个错误。然而，不像 n-gram 标注，它不计数观察结果，只编制一个转换修正规则列表。让我们看看下面的例子。

The President said he will ask Congress to increase grants to states for vocational rehabilitation.

下面将研究两个规则：

（1）当前面的词是 TO 时，替换 NN 为 VB；

（2）当下一个标记是 NNS 时，替换 TO 为 IN。

首先使用 unigram 标注器标注，然后运用规则修正错误。在此表中，我们看到两个规则，所有这些规则由以下形式产生：在上下文 C 中替换 T1 为 T2。典型的上下文是之前或之后的词的内容或标记，或者当前词 2~3 个词范围内出现的一个特定标记，在其训练阶段，T1，T2 和 C 的标注器猜测值创造出数以千计的候选规则，每一条规则根据其净收益打分，它修正不正确标记的数目减去它错误修改正确标记的数目。

Brill 标注器的另一个特性，规则是语言学可解释的，与 n-gram 标注器相比，不能从直接观察这样的一个表中学到多少东西，而 Brill 标注器学到的规则可以。

Brill 标注器有一个 "X→Y 如果前面的词是 Z" 形式的模板集合，这些模板中的变量是创建"规则"的特定词和标记的实例，规则的得分是它纠正错误例子的数目减去正确的情况下它误报的数目。除了训练标注器，还显示了剩余的错误。

```
from nltk.corpus import brown
from nltk.tag import UnigramTagger
from nltk.tag.brill import SymmetricProximateTokensTemplate, ProximateTokensTemplate
from nltk.tag.brill import ProximateTagsRule, ProximateWordsRule, FastBrillTaggerTrainer

brown_train = list(brown.tagged_sents(categories='news')[:500])
brown_test = list(brown.tagged_sents(categories='news')[500:600])
unigram_tagger = UnigramTagger(brown_train)
templates = [
SymmetricProximateTokensTemplate(ProximateTagsRule, (1, 1)),
SymmetricProximateTokensTemplate(ProximateTagsRule, (2, 2)),
```

```
    SymmetricProximateTokensTemplate(ProximateTagsRule, (1, 2)),
    SymmetricProximateTokensTemplate(ProximateTagsRule, (1, 3)),
    SymmetricProximateTokensTemplate(ProximateWordsRule, (1, 1)),
    SymmetricProximateTokensTemplate(ProximateWordsRule, (2, 2)),
    SymmetricProximateTokensTemplate(ProximateWordsRule, (1, 2)),
    SymmetricProximateTokensTemplate(ProximateWordsRule, (1, 3)),
    ProximateTokensTemplate(ProximateTagsRule, (-1, -1), (1, 1)),
    ProximateTokensTemplate(ProximateWordsRule, (-1, -1), (1, 1)),
    ]
    trainer = FastBrillTaggerTrainer(initial_tagger=unigram_tagger,
templates=templates, trace=3, deterministic=True)
    brill_tagger = trainer.train(brown_train, max_rules=10)
    brill_tagger.evaluate(brown_test) #0.742
>>> import nltk
>>> nltk.tag.brill.demo()
Loading tagged data...
Done loading.
Training unigram tagger:
    [accuracy: 0.832151]
Training bigram tagger:
    [accuracy: 0.837930]
Training Brill tagger on 1600 sentences...
Finding initial useful rules...
    Found 9757 useful rules.

           B    |
   S   F  r  O  |       Score = Fixed - Broken
   c   i  o  t  | R     Fixed = num tags changed incorrect → correct
   o   x  k  h  | u     Broken = num tags changed correct → incorrect
   r   e  e  e  | l     Other = num tags changed incorrect → incorrect
   e   d  n  r  | e
  ----------------+---------------------------------------------
   11  15  4  0 | WDT → IN if the tag of words i+1...i+2 is 'DT'
   10  12  2  0 | IN → RB if the text of the following word is
                |   'well'
    9   9  0  0 | WDT → IN if the tag of the preceding word is
                |   'NN', and the tag of the following word is 'NNP'
    7   9  2  0 | RBR → JJR if the tag of words i+1...i+2 is 'NNS'
```

```
  7 10   3   0 | WDT → IN if the tag of words i+1...i+2 is 'NNS'
  5  5   0   0 | WDT → IN if the tag of the preceding word is
              |     'NN', and the tag of the following word is 'PRP'
  4  4   0   1 | WDT → IN if the tag of words i+1...i+3 is 'VBG'
  3  3   0   0 | RB → IN if the tag of the preceding word is 'NN',
              |     and the tag of the following word is 'DT'
  3  3   0   0 | RBR → JJR if the tag of the following word is
              |     'NN'
  3  3   0   0 | VBP → VB if the tag of words i-3...i-1 is 'MD'
  3  3   0   0 | NNS → NN if the text of the preceding word is
              |     'one'
  3  3   0   0 | RP → RB if the text of words i-3...i-1 is 'were'
  3  3   0   0 | VBP → VB if the text of words i-2...i-1 is "n't"

Brill accuracy: 0.839156
Done; rules and errors saved to rules.yaml and errors.out.
>>>
```

5.7 确定一个词的分类

5.7.1 形态学线索

一个词的内部结构可能为这个词分类提供有用的线索。举例来说，-ness 是一个后缀，与形容词结合产生一个名词，如 happy→happiness，ill→illness。因此，如果遇到一个以-ness 结尾的词，很可能是一个名词。同样地，-ment 是与一些动词结合产生一个名词的后缀，如 govern→government 和 establish→establishment。英语动词也可以是形态复杂的，例如，一个动词的现在分词以-ing 结尾，表示正在进行的还没有结束的行为，如 falling。-ing 后缀也出现在动名词中，如 the falling of the leaves。

5.7.2 句法线索

另一个信息来源是一个词可能出现的上下文语境，例如，假设我们已经确定了名词类，那么可以说，英语形容词的句法标准是它可以立即出现在一个名词前，或紧跟在词 be 或 very 后。根据这些测试，near 应该被归类为形容词。

```
a. the near window
b. The end is very near.
```

5.7.3 语义线索

一个词的语义对其词性是一个有用的线索。例如，名词的语义是"一个人、地方或事物的名称。"在现代语言学，词类的语义标准受到怀疑，主要是因为它们很难规范化。然而，语义标准巩固了我们对许多词类的直觉，使我们能够在不熟悉的语言中很好地猜测词的分类。例如，如果我们都知道荷兰语 verjaardag 的意思与英语 birthday 相同，那么可以猜测 verjaardag 在荷兰语中是一个名词。然而，一些修补是必要的，虽然可能 zij is vandaag jarig 将翻译为 It's her birthday today。词 jarig 在荷兰语中实际上是形容词，与英语并不完全相同。

小 结

（1）词可以组成类，如名词、动词、形容词及副词。这些类被称为词汇范畴或者词性。词性被分配短标签或者标记，如 NN 和 VB。

（2）给文本中的词自动分配词性的过程称为词性标注、POS 标注或标注。

（3）自动标注是 NLP 流程中重要的一步，在各种情况下都十分有用，包括预测先前未见过的词的行为、分析语料库中词的使用以及文本到语音转换系统。

（4）一些语言学语料库，如 brown 语料库，已经做了词性标注。

（5）有多种标注方法，如默认标注器、正则表达式标注器、unigram 标注器、n-gram 标注器。这些都可以结合一种叫作回退的技术一起使用。

（6）标注器可以使用已标注语料库进行训练和评估。

（7）回退是一个组合模型的方法：当一个较专业的模型（如 bigram 标注器）不能为给定内容分配标记时，可以回退到一个较一般的模型（如 unigram 标注器）。

（8）词性标注是 NLP 中一个重要的早期的序列分类任务：利用局部上下文语境中的词和标记对序列中任意一点的分类决策。

（9）字典用来映射任意类型之间的信息，如字符串和数字：freq['cat']=12。使用大括号来创建字典：pos={}, pos={'furiously':'adv', 'ideas':'n', 'colorless':'adj'}。

（10）N-gram 标注器可以定义较大数值的 n，但是当 n 大于 3 时，我们常常会面临数据稀疏问题，即使使用大量的训练数据，看到的也只是可能的上下文的一小部分。

（11）基于转换的标注学习一系列"改变标记 s 为标记 t 在上下文 c 中"形式的修复规则，每个规则会修复错误，也可能引入（较小的）错误。

练 习

1. 网上搜索"spoof newspaper headlines"，找到 British Left on Falkland Islands 和 Juvenile Court to Try Shooting Defendant，手工标注这些词条，看看词性标记的知识是否可以消除歧义。

```
Please bo my flight for Delhi.
  [('Please', 'V'), ('bo', 'V'), ('my', 'PRO'), ('flight', 'N'),
('for', 'P'), ('Delhi', 'NP')]
```

简化标记 V=VB，P=IN，PRO=pronoun?PRP。

2. 和别人一起，轮流挑选一个既可以是名词也可以是动词的词，如 contest，refuse，bo，permit，让对方预测哪一个可能是 brown 语料库中频率最高的。检查对方的预测，为几个回合打分。

```
from nltk import ConditionalFreqDist
from nltk.corpus import brown
tagged_words=brown.tagged_words()
cfd=ConditionalFreqDist(tagged_words)
cfd['contest'].max() #'NN'
contest 英[ˈkɒntest，kənˈtest]美[ˈkɑːntest，kənˈtest]
n. 竞赛；比赛；(控制权或权力的)争夺，竞争；
vt. 争取赢得(比赛、选举等)；争辩；就…提出异议；
```

3. 分词和标注下面的句子：They wind back the clock，while we chase after the wind。它包含哪些不同的发音和词类？

```
import nltk
raw='They wind back the clock, while we chase after the wind'
sent=nltk.word_tenize(raw)
tagged_sent=nltk.pos_tag(sent)
 [('They', 'PRP'), ('wind', 'VBP'), ('back', 'RB'), ('the', 'DT'),
('clock', 'NN'), (',', ','), ('while', 'IN'), ('we', 'PRP'), ('chase',
'VBP'), ('after', 'IN')
, ('the', 'DT'), ('wind', 'NN')]
 n. [wɪnd]
 v. [waɪnd]
```

4. 回顾表 5-4 中的映射。讨论你能想到的映射的其他例子。它们是从什么类型的信息映射到什么类型的信息？

（1）文档索引：页面列表（找到词的地方）=f（词）；

（2）同义词：同义词列表=f（词义）；

（3）词典：词条项（词性、意思定义、词源）=f（中心词）；

（4）比较单词列表：同源词（词列表，每种语言一个）=f（注释术语）；

（5）词形分析：形态学分析（词素组件列表）=f（表面形式）；

（6）词性标注：标记=f（词）；

（7）电话簿：数字=f（名字）；

（8）域名解析：IP 地址=f（域名）；

（9）词频表：频率=f（词）。

5. 在交互模式下使用 Python 解释器，实验本章中字典的例子。创建一个字典 d，添加一些条目。尝试访问一个不存在的条目，如 d['xyz']，看看会发生什么？

```
d={}
d['Excuse']=1
d['me']=1
d['xyz']
d={'Is':1, 'there':1}
d=dict(any=1, summer=1)
d=dict([('resort', 1), ('around', 1)])
strings=('puppy', 'kitten', 'puppy', 'puppy',
'weasel', 'puppy', 'kitten', 'puppy')
counts={}
for kw in set(strings):
    counts[kw]=0

for kw in strings:
    counts[kw]+=1
```

6. 尝试从字典 d 删除一个元素，使用语法 del d['abc']，检查被删除的项目。

```
d={}
d['abc']=1
d['efg']=1
del d['abc']
```

7. 创建两个字典 d1 和 d2，为每个添加一些条目。发出命令 d1.update(d2)，它会做什么？可能有什么用？

```
d1={'Excuse':1, 'me':1}
d2={'Is':1, 'there':1}
d1.update(d2)
```

8. 创建一个字典 e，表示选择的一些词的一个单独的词汇条目。定义键，如 headword、part-of-speech、sense 和 example，并分配适当的值。

```
e=dict(headword='file',   part_of_speech='NN',   sense='fail',
example='The file is printable .')
e={'headword':'file',   'part-of-speech':'NN',   'sense':'fail',
'example':'The file is printable .'}
```

9. 自己验证 go 和 went 在分布上的限制。

```
from nltk.bo import *
```

```
text1.concordance('go')
text1.concordance('went')
```

10. 训练一个 unigram 标注器，在一些新的文本上运行。观察到有些词没有分配到标记，这是为什么？

```
from nltk.corpus import brown
brown_tagged_sents=brown.tagged_sents(categories='news')
brown_sents=brown.sents(categories='news')
unigram_tagger=nltk.UnigramTagger(brown_tagged_sents)
unigram_tagger.tag(brown_sents[2007])
unigram_tagger.evaluate(brown_tagged_sents)
#0.9349006503968017
[('Various', 'JJ'), ('of', 'IN'), ('the', 'AT'), ('apartments',
'NNS'), ('are', 'BER'), ('of', 'IN'), ('the', 'AT'), ('terrace', 'NN'),
('type', 'NN'), (',', ','), ('being', 'BEG'), ('on', 'IN'), ('the',
'AT'), ('ground', 'NN'), ('floor', 'NN'), ('so', 'QL'), ('that', 'CS'),
('entrance', 'NN'), ('is', 'BEZ'), ('direct', 'JJ'), ('.', '.')]
```

11. 了解词缀标注器（输入 help(nltk.AffixTagger)）。训练一个词缀标注器，在一些新的文本上运行。设置不同的词缀长度和最小词长做实验，讨论结果。

```
import nltk
from nltk.corpus import brown
brown_tagged_sents=brown.tagged_sents(categories='news')
brown_sents=brown.sents(categories='news')
affixtagger=nltk.AffixTagger(brown_tagged_sents)
affixtagger.tag(brown_sents[2007])
affixtagger.evaluate(brown_tagged_sents)
[('Various', 'JJ'), ('of', None), ('the', None), ('apartments',
'NNS'), ('are', ne), ('of', None), ('the', None), ('terrace', 'NN'),
('type', None), (',', Non, ('being', 'VBG'), ('on', None), ('the', None),
('ground', 'NN'), ('floor', ''), ('so', None), ('that', None),
('entrance', 'NN'), ('is', None), ('direct', NN'), ('.', None)]
0.25591224615629415
```

12. 训练一个没有回退标注器的 bigram 标注器，在一些训练数据上运行。然后，在一些新的数据上运行它。标注器的性能会发生什么？为什么呢？

```
import nltk
from nltk.corpus import brown
brown_tagged_sents=brown.tagged_sents(categories='news')  #否则很慢
```

```
brown_sents=brown.sents(categories='news')
size=int(len(brown_tagged_sents)*0.9)
train_sents=brown_tagged_sents[:size]
test_sents=brown_tagged_sents[size:]
bigram_tagger=nltk.BigramTagger(train_sents)
bigram_tagger.tag(brown_sents[2007])
bigram_tagger.tag(brown_sents[4160])
bigram_tagger.evaluate(test_sents) #0.10216286255357321, 降低, None
 [('Various', 'JJ'), ('of', 'IN'), ('the', 'AT'), ('apartments',
'NNS'), ('are', 'BER'), ('of', 'IN'), ('the', 'AT'), ('terrace', 'NN'),
('type', 'NN'), (',', ','), ('being', 'BEG'), ('on', 'IN'), ('the',
'AT'), ('ground', 'NN'), ('floor', 'NN'), ('so', 'CS'), ('that', 'CS'),
('entrance', 'NN'), ('is', 'BEZ'), ('direct', 'JJ'), ('.', '.')]
 [('But', 'CC'), ('in', 'IN'), ('all', 'ABN'), ('its', 'PP$'), ('175',
None), ('years', None), (',', None), ('not', None), ('a', None),
('single', None), ('student', None), ('has', None), ('entered', None),
('its', None), ('classrooms', None), ('.', None)]
```

13. 可以使用字典指定由一个格式化字符串替换的值。阅读关于格式化字符串的 Python 库文档，使用这种方法以两种不同的格式显示今天的日期。

```
d={'year':2020, 'month':1, 'day':1}
print '%(year)s-%(month)s-%(day)s'%d
print '%s-%s-%s'%(d['year'], d['month'], d['day'])
```

14. 使用 sorted()和 set()获得 brown 语料库使用的标记的排序的列表，删除重复。

```
from nltk.corpus import brown
brown_tagged_words = brown.tagged_words(simplify_tags=True)
tags=[tag for _, tag in brown_tagged_words]
sorted(set(tags))
['', "'", "'", '(', ')', '*', ',', '.', ':', 'ADJ', 'ADV', 'CNJ',
'DET', 'EX', 'FW', 'MOD', 'N', 'NP', 'NUM', 'P', 'PRO', 'TO', 'UH', 'V',
'VB+PPO', 'VBZ', 'VD', 'VG', 'VN', 'WH', '``']
```

15. 写程序处理 brown 语料库，找到以下问题的答案。

（1）哪些名词常以它们复数形式而不是它们的单数形式出现（只考虑常规的复数形式，-s 后缀形式，NNS: noun, plural, common, NN: noun, singular, common）？

```
import nltk
from nltk.corpus import brown
brown_tagged_words=brown.tagged_words()
```

```
    singular=[word.lower() for word, tag in brown_tagged_words if
tag=='NN']
    fd1=nltk.FreqDist(singular)
    plural=[word.lower() for word, tag in brown_tagged_words if tag=='NNS']
    fd2=nltk.FreqDist(plural)
    [w for w in fd1 if w+'s' in fd2 and fd2[w+'s']/fd1[w]>10]
    ['stair', 'stockholder', 'tear', 'employe', 'headquarter',
'investor', 'microorg
    anism', 'relative', 'rib', 'survivor']
```

（2）哪些词的不同标记数目最多？它们是什么？它们代表什么？

```
    import nltk
    from nltk.corpus import brown
    brown_tagged_words=brown.tagged_words()
    cfd=nltk.ConditionalFreqDist(brown_tagged_words)
    y=max([(len(cfd[x]), x) for x in cfd])
    y[1]
    cfd[y[1]].keys()
    'that'
    ['CS', 'DT', 'WPS', 'WPO', 'QL', 'DT-NC', 'WPS-NC', 'CS-NC', 'WPS-HL',
'CS-HL', 'NIL', 'WPO-NC']
    CS: conjunction, subordinating
    nltk.help.brown_tagset('CS.*')
```

（3）按频率递减的顺序列出标记。前 20 个最频繁的标记代表什么？

```
    import nltk
    from nltk.corpus import brown
    from collections import Counter
    brown_tagged_words=brown.tagged_words()
    tags=[tag for _, tag in brown_tagged_words]
    fd=nltk.FreqDist(tags)
    [x for x, _ in Counter(fd).most_common(20)]
    ['NN', 'IN', 'AT', 'JJ', '.', ',', 'NNS', 'CC', 'RB', 'NP', 'VB', 'VBN', 'VBD',
'CS', 'PPS', 'VBG', 'PP$', 'TO', 'PPSS', 'CD']
```

（4）名词后面最常见的是哪些标记？这些标记代表什么？

```
    import nltk
    from nltk.corpus import brown
    brown_tagged_words=brown.tagged_words()
```

```
tags=[y[1] for x, y in nltk.bigrams(brown_tagged_words) if x[1]=='NN']
fd=nltk.FreqDist(tags)
fd.max()
'IN'
```

16. 探索回答有关查找标注器的以下问题。

（1）回退标注器被省略时，模型大小变化，标注器的性能会发生什么？

随着模型规模的增长，最初的性能增加迅速，最终达到一个稳定水平，这时模型的规模大量增加，但性能的提高却很小。

（2）思考曲线，为查找标注器推荐一个平衡内存和性能较好的规模。在什么情况下应该尽量减少内存使用？在什么情况下性能最大化而不必考虑内存使用？

size=2048，如果使用各种语言技术的标注器部署在移动计算设备上，在模型大小和标注器性能之间取得平衡是很重要的。

17. 查找标注器的性能上限是什么？假设其表的大小没有限制（提示：写一个程序算出被分配了最有可能的标记的词的标识符的平均百分比）。

```
import nltk
from nltk.corpus import brown
from __future__ import division
brown_tagged_sents=brown.tagged_sents(categories='news')
cfd=nltk.ConditionalFreqDist(x for sent in brown_tagged_sents for x in sent)
ambiguous_contexts=[c for c in cfd.conditions() if len(cfd[c])>1]
print sum(cfd[c].N() for c in ambiguous_contexts)/cfd.N()
#0.580842134574≈58%
```

18. 生成已标注数据的一些统计数据，回答下列问题。

（1）总是被分配相同词性的词类的比例是多少？

```
import nltk
from nltk.corpus import brown
from nltk import ConditionalFreqDist
from __future__ import division
brown_tagged_sent= [('They', 'PRP'), ('refuse', 'VBP'), ('refuse', 'NN')]
cfd=ConditionalFreqDist(brown_tagged_sent)
x=0
for c in cfd.conditions():
    if len(cfd[c].keys())==1:
        x+=1

x/len(cfd.keys())  #0.5
```

（2）多少词是有歧义的（即从某种意义上说，它们至少和两个标记一起出现）？

（3）brown语料库中这些有歧义的词的标识符的百分比是多少？

19. evaluate()方法算出一个文本上运行的标注器的精度。例如，如果提供的已标注文本是[('the', 'DT'), ('dog', 'NN')]，标注器产生的输出是[('the', 'NN'), ('dog', 'NN')]，那么得分为0.5。尝试找出评价方法是如何工作的。

（1）一个标注器t将一个词汇列表作为输入，产生一个已标注词列表作为输出。t.evaluate()只以一个正确标注的文本作为唯一的参数。执行标注之前必须对输入做些什么？

```
from nltk import pos_tag
from nltk.tag.util import untag
from nltk.metrics import accuracy
gold = [[('the', 'DT'), ('dog', 'NN')]]
sents = [untag(sent) for sent in gold]
sentences = sents
tagged_sents = [pos_tag(sent) for sent in sentences] #?
gold_tens = sum(gold, [])
test_tens = sum(tagged_sents, [])
accuracy(gold_tens, test_tens)
```

（2）一旦标注器创建了新标注的文本，evaluate()方法可能如何比较它与原来标注的文本，计算准确性得分。

```
from nltk import pos_tag
from nltk.tag.util import untag
from nltk.metrics import accuracy
gold = [[('the', 'DT'), ('dog', 'NN')]]
sents = [untag(sent) for sent in gold]
sentences = sents
tagged_sents = [pos_tag(sent) for sent in sentences] #?
gold_tens = sum(gold, [])
test_tens = sum(tagged_sents, [])
accuracy(gold_tens, test_tens)
```

（3）现在，检查源代码来看看这个方法是如何实现的。检查nltk.tag.api.__file__找到源代码的位置，使用编辑器打开这个文件（一定要使用文件api.py，而不是编译过的二进制文件api.pyc）。

```
from nltk import pos_tag
from nltk.tag.util import untag
from nltk.metrics import accuracy
gold = [[('the', 'DT'), ('dog', 'NN')]]
sents = [untag(sent) for sent in gold]
sentences = sents
```

```
tagged_sents = [pos_tag(sent) for sent in sentences] #?
gold_tens = sum(gold, [])
test_tens = sum(tagged_sents, [])
accuracy(gold_tens, test_tens)
```

20. 编写代码，搜索 brown 语料库，根据标记查找特定的词和短语，回答下列问题。
（1）产生一个标注为 MD 的不同的词的按字母顺序排序的列表。
（2）识别可能是复数名词或第三人称单数动词的词（如 deals，flies）。
（3）识别三个词的介词短语形式 IN+DT+NN（如 in the lab）。

```
import nltk
raw='Excuse me . Is there any summer resort around the town ?'
words=nltk.word_tenize(raw)
sent=nltk.pos_tag(words)
[' '.join([w1,w2,w3]) for (w1,t1),(w2,t2),(w3,t3) in nltk.trigrams(sent)
if t1=='IN' and t2=='DT' and t3=='NN']
for (w1,t1),(w2,t2),(w3,t3) in nltk.trigrams(sent):
    if t1=='IN' and t2=='DT' and t3=='NN':
        print w1,w2,w3
```

（4）男性与女性代词的比例是多少？

```
import nltk
from nltk.corpus import brown
brown_words=brown.words()
male_words=[x for x in brown_words if x in {'he','him','his','himself'}]
female_words=[x for x in brown_words if x in {'she','her','hers','herself'}]
male_len=len(male_words)
female_len=len(female_words)
brown_len=len(brown_words)
male_ratio=male_len*1.0/brown_len*100
female_ratio=female_len*1.0/brown_len*100
male_ratio
female_ratio
```

21. 定义可以用来做生词的回退标注器的 regexp_tagger。这个标注器只检查基数词。通过特定的前缀或后缀字符串进行测试，它应该能够猜测其他标记。例如，可以标注所有 -s 结尾的词为复数名词。定义一个正则表达式标注器（使用 RegexpTagger()），测试至少 5 个单词拼写的其他模式（使用内联文档解释规则）。

22. 考虑上一练习中开发的正则表达式标注器，使用它的 accuracy() 方法评估标注器，尝试想办法提高其性能。讨论你找到的结果。客观的评估如何帮助开发过程？

23. 数据稀疏问题有多严重？测试 n-gram 标注器当 n 从 1 增加到 6 时的性能。为准确性得分制表。估计这些标注器需要的训练数据，假设词汇量大小为 105 而标记集的大小为 102。

24. 获取另一种语言的一些已标注数据，在其上测试和评估各种标注器。如果这种语言是形态复杂的，或者有词类的任何字形线索（如 capitalization），可以考虑为它开发一个正则表达式标注器（排在 unigram 标注器之后，默认标注器之前）。对比同样的运行在英文数据上的标注器，你的 tagger(s)的准确性如何？讨论你在运用这些方法到这种语言时遇到的问题。

25. 绘制曲线显示查找标注器的性能随模型大小增加的变化。绘制训练数据量变化时 unigram 标注器的性能曲线。

26. 检查 5.5 节中定义的 bigram 标注器 t2 的混淆矩阵，确定简化的一套或多套标记。定义字典做映射，在简化的数据上评估标注器。

27. 使用简化的标记集测试标注器（或制作一个你自己的，通过丢弃每个标记名中除第一个字母外所有的字母）。这种标注器需要做的区分更少，但由它获得的信息也更少，讨论结果。

28. 回顾一个 bigram 标注器训练过程中遇到生词，标注句子的其余部分为 None 的例子。一个 bigram 标注器只处理了句子的一部分就失败了，即使句子中没有包含生词（即使句子在训练过程中使用过），这可能吗？在什么情况下会出现这种情况呢？你可以写一个程序，找到一些这方面的例子吗？

29. 预处理布朗新闻数据，替换低频词为 UNK，但留下标记不变。在这些数据上训练和评估一个 bigram 标注器。这样有多少帮助？unigram 标注器和默认标注器的贡献是什么？

```
import nltk
from nltk.corpus import brown
brown_tagged_sents=brown.tagged_sents(categories='news')
news1=brown.words(categories='news')
vocab=nltk.FreqDist(news1)
v500=list(vocab)[:500]
mapping=nltk.defaultdict(lambda:'UNK')
for v in v500:
    mapping[v]=v

news2=[]
for x in brown_tagged_sents:
    news2.append([(mapping[v[0]], v[1]) for v in x])

news2[0]
import nltk
from nltk.corpus import brown
brown_tagged_sents=brown.tagged_sents(categories='news')
brown_sents=brown.sents(categories='news')
size=int(len(brown_tagged_sents)*0.9)  #4160
```

```
train_sents=brown_tagged_sents[:size]
test_sents=brown_tagged_sents[size:]
bigram_tagger=nltk.BigramTagger(train_sents)
bigram_tagger.tag(brown_sents[2007])
unseen_sent=brown_sents[4203] #4160~4623
bigram_tagger.tag(unseen_sent)
bigram_tagger.evaluate(test_sents) #0.10216286255357321
```
[('Various', 'JJ'), ('of', 'IN'), ('the', 'AT'), ('apartments', 'NNS'), ('are', 'BER'), ('of', 'IN'), ('the', 'AT'), ('terrace', 'NN'), ('type', 'NN'), (',', ','), ('being', 'BEG'), ('on', 'IN'), ('the', 'AT'), ('ground', 'NN'), ('floor', 'NN'), ('so', 'CS'), ('that', 'CS'), ('entrance', 'NN'), ('is', 'BEZ'), ('direct', 'JJ'), ('.', '.')]

[('The', 'AT'), ('population', 'NN'), ('of', 'IN'), ('the', 'AT'), ('Congo', 'NP'), ('is', 'BEZ'), ('13.5', None), ('million', None), (',', None), ('divided', None), ('into', None), ('at', None), ('least', None), ('seven', None), ('major', None), ('``', None), ('culture', None), ('clusters', None), ("''", None), ('and', None), ('innumerable', None), ('tribes', None), ('speaking', None), ('400', None), ('separate', None), ('dialects', None), ('.', None)]

```
import nltk
from nltk.corpus import brown
brown_tagged_sents=brown.tagged_sents(categories='news')
brown_sents=brown.sents(categories='news')
brown_words=brown.words(categories='news')
vocab=nltk.FreqDist(brown_words)
v1000=list(vocab)[:100554]
mapping=nltk.defaultdict(lambda:'UNK')
for v in v1000:
    mapping[v]=v

news1=[]
for x in brown_sents:
    news1.append([mapping[v] for v in x])

news2=[]
for x in brown_tagged_sents:
    news2.append([(mapping[v[0]], v[1]) for v in x])

brown_tagged_sents=news2
```

```
brown_sents=news1

size=int(len(brown_tagged_sents)*0.91) #4160
train_sents=brown_tagged_sents[:size]
test_sents=brown_tagged_sents[size:]
bigram_tagger=nltk.BigramTagger(train_sents)
bigram_tagger.tag(brown_sents[2007])
unseen_sent=brown_sents[4203] #4160~4623
bigram_tagger.tag(unseen_sent)
bigram_tagger.evaluate(test_sents) #0.35303498455098176
```

30. 修改程序，通过将 pylab.plot()替换为 pylab.semilogx()，在 X 轴上使用对数刻度。关于结果图形的形状，你注意到了什么？梯度告诉你什么呢？

31. 阅读 Brill 标注器演示函数的文档，使用 help(nltk.tag.brill.demo)。通过设置不同的参数值试验这个标注器。是否有训练时间（语料库大小）和性能之间的权衡？

```
C:\Python27\Lib\site-packages\nltk\tag\brill.py
```

（1）将 brill.py 拷贝到 C:\Python27；
（2）>>> import brill；
（3）>>> brill.demo()。

```
from nltk.corpus import brown
from nltk.tag import UnigramTagger
from nltk.tag.brill import SymmetricProximateTensTemplate, ProximateTensTemplate
from nltk.tag.brill import ProximateTagsRule, ProximateWordsRule, FastBrillTaggerTrainer
brown_train = list(brown.tagged_sents(categories='news')[:500])
brown_test = list(brown.tagged_sents(categories='news')[500:600])
unigram_tagger = UnigramTagger(brown_train)
templates = [
SymmetricProximateTensTemplate(ProximateTagsRule, (1, 1)),
SymmetricProximateTensTemplate(ProximateTagsRule, (2, 2)),
SymmetricProximateTensTemplate(ProximateTagsRule, (1, 2)),
SymmetricProximateTensTemplate(ProximateTagsRule, (1, 3)),
SymmetricProximateTensTemplate(ProximateWordsRule, (1, 1)),
SymmetricProximateTensTemplate(ProximateWordsRule, (2, 2)),
SymmetricProximateTensTemplate(ProximateWordsRule, (1, 2)),
SymmetricProximateTensTemplate(ProximateWordsRule, (1, 3)),
ProximateTensTemplate(ProximateTagsRule, (-1, -1), (1, 1)),
```

```
ProximateTensTemplate(ProximateWordsRule, (-1, -1), (1, 1)),
]
trainer = FastBrillTaggerTrainer(initial_tagger=unigram_tagger,
 templates=templates, trace=3,
 deterministic=True)
brill_tagger = trainer.train(brown_train, max_rules=10)
brill_tagger.evaluate(brown_test)  #0.7427832830676433
```

32. 编写代码构建一个集合的字典。用它来存储一套可以跟在具有给定 POS 标记的给定词后面的 POS 标记，如 word_i→tag_i→tag_{i+1}。

33. brown 语料库中有 264 个不同的词有 3/4 种可能的标签。

（1）打印一个表格，一列是整数 1～10，另一列是语料库中有 1～10 个不同标记的不同词的数目。

```
import nltk
from nltk.corpus import brown
btws=brown.tagged_words()
cfd=nltk.ConditionalFreqDist(btws)
cl=[len(cfd[c]) for c in cfd.conditions()]
fd=nltk.FreqDist(cl)
for x in fd.items():
    if x[0]<=10:
        print '%-4s%-8s'%(x[0], x[1])
1    47328 有 1 个不同标记的不同词的数目是 47328
2    7186
3    1146 有 3 个不同标记的不同词的数目是 1146
4    265
5    87
6    27
7    12
8    1
9    1
10   2
```

（2）对有不同的标记数量最多的词，输出语料库中包含这个词的句子，每种可能的标记一个。

```
import nltk
from nltk.corpus import brown
btws=brown.tagged_words()
cfd=nltk.ConditionalFreqDist(btws)
```

```
cl=[len(cfd[c]) for c in cfd.conditions()]

m=0
for c in cfd.conditions():
    if len(cfd[c])>m:
        m=len(cfd[c])
        t=c

bss=brown.sents()
i=0
l=[]
for bs in bss:
    if t in bs:
        l.append((i, bs.index(t)))
    i=i+1

btss=brown.tagged_sents()
y=[]
for x in l:
    if btss[x[0]][x[1]] not in y:
        btss[x[0]]
        y.append(btss[x[0]][x[1]])
```

34. 编写一个程序，按照词 must 后面的词的标记为它的上下文分类。这样可以区分 must 的 "必须" 和 "应该" 两种词义上的用法吗？

```
import nltk
from nltk.corpus import brown
ws=brown.tagged_words()
bs=nltk.bigrams(ws)
ts=[]
for b in bs:
    if b[0][0]=='must':
        ts.append(b[1][1])
fd=nltk.FreqDist(ts)
FreqDist({u'BE': 369, u'VB': 353, u'HV': 134, u'RB': 74, u'*': 21, u',': 17, u'DO': 7, u'IN': 6, u"'": 3, u'PPO': 2, ...})
```

35. 创建一个正则表达式标注器和各种 unigram 以及 n-gram 标注器（包括回退），在 brown 语料库上训练它们。

（1）创建这些标注器的 3 种不同组合。测试每个组合标注器的准确性。哪种组合效果最好？

```
import nltk
from nltk.corpus import brown
brown_tagged_sents=brown.tagged_sents(categories='news')
brown_sents=brown.sents(categories='news')
patterns=[
(r'.*ing$', 'VBG'), #gerunds 动名词
(r'.*ed$', 'VBD'), #simple past
(r'.*es$', 'VBZ'), #3rd singular present
(r'.*ould$', 'MD'), #modals
(r'.*\'s$', 'NN$'), #possessive nouns
(r'.*s$', 'NNS'), #plural nouns
(r'^-?[0-9]+(.[0-9]+)?$', 'CD'), #cardinal numbers
(r'.*', 'NN') #nouns(default)
]
regexp_tagger=nltk.RegexpTagger(patterns) #不需要训练！
regexp_tagger.tag(brown_sents[3])
regexp_tagger.evaluate(brown_tagged_sents)
#0.20326391789486245=1/5
import nltk
from nltk.corpus import brown
tagged_sents=brown.tagged_sents(categories='news')
size=int(len(tagged_sents)*0.9)
train_sets=tagged_sents[:size]
test_sets=tagged_sents[size:]
t0=nltk.DefaultTagger('NN')
t1=nltk.UnigramTagger(train=train_sets, backoff=t0)
t2=nltk.BigramTagger(train=train_sets, backoff=t1)
x=t2.tag(train_sets[0])
print t2.evaluate([x]) #1.0
print t2.evaluate(train_sets) #0.973597286817
print t2.evaluate(test_sets) #0.844911791089
```

（2）尝试改变训练语料的规模。它是如何影响结果的？

36. 标注生词的方法一直要考虑这个词的字母（使用 RegexpTagger()），或完全忽略这个词，将它标注为一个名词（使用 nltk.DefaultTagger()）。这些方法对于有新词却不是名词的文本不太友好。思考句子：I like to blog on Kim's blog，如果 blog 是一个新词，查看前面的标记（TO 和 NP$）。

我们需要一个对前面的标记敏感的默认标注器。

（1）创建一种新的 unigram 标注器，查看前一个词的标记，而忽略当前词。（做到这一点的最好方法是修改 UnigramTagger()的源代码，需要用到 Python 中面向对象编程的知识）

（2）将这个标注器加入回退标注器序列（包括普通的 trigram 和 bigram 标注器），放在常用默认标注器的前面。

（3）评价这个新的 unigram 标注器的作用。

37. 思考 5.5 节中的代码，它确定一个 trigram 标注器的准确性上限。回顾 Abney 关于精确标注的不可能性的讨论，解释为什么正确标注这些例子需要获取词和标记以外的其他种类的信息。你如何估计这个问题的规模？

38. 使用 nltk.probability 中的一些估计技术，如 Lidstone 或 Laplace 估计，开发一种统计标注器，它在训练中没有遇到而测试中遇到的上下文中表现优于 n-gram 回退标注器。

> probability [ˌprɒbəˈbɪləti]
> n. 可能性；或然性；很可能发生的事；概率；几率；或然率；
> [例句]Without a transfusion, the victim's probability of dying was 100%

39. 开发一个 n-gram 回退标注器，允许在标注器初始化时指定"anti-n-grams"，如["the", "the"]。一个 anti-n-grams 被分配一个数字 0，被用来防止这个 n-gram 回退（如避免估计 P(the|the) 而只做 P(the)）。

40. 使用 brown 语料库开发标注器时，测试三种不同的方式来定义训练和测试数据之间的分割：genre(category)、source(fileid)和句子。比较它们的相对性能，并讨论哪种方法最合理（可能要使用 n-交叉验证，以提高评估的准确性）。

41. 开发你自己的 NgramTagger，从 NLTK 中的类继承，封装本章中所述的已标注的训练和测试数据的词汇表缩减方法，确保 unigram 和默认回退标注器有机会获得全部词汇。

第 6 章 文本分类

6.1 有监督分类

分类是为给定的输入选择正确类标签。在基本的分类中，每个输入认为是与所有其他输入无关的（独立的），并且标签集是预先定义的。

如果分类是基于每个输入正确标签的训练语料，称为有监督分类。在训练过程中，特征提取器用来将每一个输入值转换为特征集，这些特征集捕捉每个输入中用于对其分类的基本信息，特征集与标签的配对送入机器学习算法，生成模型。在预测过程中，相同的特征提取器用来将未见过的输入转换为特征集，然后，这些特征集送入模型产生预测标签。

6.1.1 性别分类

男性和女性的名字有一些鲜明的特点，以 a, e 和 i 结尾的很可能是女性，而以 k, o, r, s 结尾的很可能是男性，试建立一个分类器更精确地模拟这些差异。

第一步，决定输入的什么特征是相关的以及如何为那些特征编码，在这个例子中，开始只是寻找一个给定名字的最后一个字母。以下特征提取器建立一个字典，包含有关给定名字的相关信息。

```
>>> def gender_features(n):
...     return {'last_letter':n[-1]}
...
>>> gender_features('Kate')
{'last_letter': 'e'}
```

这个函数返回的字典{'last_letter': 'e'}被称为特征集，映射特征名称到它们的值。特征名称是区分大小写的字符串，通常用一个简短的特征描述，特征值是简单类型的值，如布尔、数字和字符串，大多数分类方法要求特征使用简单的类型进行编码，如布尔类型、数字和字符串。

第二步，准备一个例子和对应类标签的列表，包含每个输入正确标签的训练语料。

```
>>> from nltk.corpus import names
>>> import random
```

```
>>> names=([(n, 'male') for n in names.words('male.txt')]+
... [(n, 'female') for n in names.words('female.txt')])
>>> random.shuffle(names)
>>> names[:3]
[('Bessy', 'female'), ('Mendie', 'male'), ('Kara-Lynn', 'female')]
```

第三步，使用特征提取器处理名称数据并划分特征集的结果列表为一个训练集和一个测试集。训练集用于训练一个新的朴素贝叶斯分类器。

```
>>> def gender_features(n):
...     return {'last_letter':n[-1]}
...
>>> from nltk.corpus import names
>>> from random import shuffle
>>> names=([(n, 'male') for n in names.words('male.txt')]+
... [(n, 'female') for n in names.words('female.txt')])
>>> shuffle(names)
>>>
>>> featuresets=[(gender_features(n), g) for n, g in names]
>>> train_set, test_set=featuresets[500:], featuresets[:500]
>>> from nltk import NaiveBayesClassifier
>>> classifier=NaiveBayesClassifier.train(train_set)
```

第四步，测试一些没有出现在训练数据中的名字。

```
>>> classifier.classify(gender_features('Tom'))
'male'
>>> classifier.classify(gender_features('Kate'))
'female'
```

显然，这些名字被正确分类，符合有关名字和性别的预期。

第五步，在大数据量的未见过的数据上系统地评估这个分类器。

```
>>> from nltk.classify import accuracy
>>> accuracy(classifier, test_set)
0.794
```

第六步，检查分类器，确定哪些特征对于区分名字的性别是最有效的。

```
>>> classifier.show_most_informative_features(5)
Most Informative Features
         last_letter = 'a'         female : male   =    34.3 : 1.0
         last_letter = 'k'         male : female   =    31.4 : 1.0
```

```
            last_letter = 'f'        male : female =     15.4 : 1.0
            last_letter = 'p'        male : female =     12.6 : 1.0
            last_letter = 'v'        male : female =     11.3 : 1.0
```

此列表显示训练集中以 a 结尾的名字中女性是男性的 34 倍，以 a 结尾的名字最有可能是女性，而以 k 结尾名字中男性是女性的 31 倍，以 a 结尾的名字最有可能是男性，这些比率称为似然比，可以用于比较不同特征-结果关系。

修改 gender_features()函数，为分类器提供名称的长度、它的第一个字母以及任何其他看起来可能有用的特征，再用这些新特征训练分类器并测试其准确性。

在处理大型语料库时，构建一个包含每一个实例特征单独的列表会使用大量的内存，在这些情况下，使用函数 nltk.classify.util.apply_features，返回一个行为像一个列表而不会在内存存储所有特征集的对象（产生器，类似于语法分析与词法分析的关系）。

```
>>> from nltk.corpus import names
>>> from random import shuffle
>>> from nltk.classify.util import apply_features
>>> from nltk import NaiveBayesClassifier
>>> from nltk.classify import accuracy
>>> names=([(name, 'male') for name in names.words('male.txt')]+
... [(name, 'female') for name in names.words('female.txt')])
>>> shuffle(names)
>>> def gender_features(name):
...         return {'last_letter': name[-1].lower(), 'length':
len(name), 'first_letter': name[0].lower()}
...
>>> train_set = apply_features(gender_features, names[500:])
>>> test_set = apply_features(gender_features, names[:500])
>>> classifier=NaiveBayesClassifier.train(train_set)
>>> classifier.classify(gender_features('Tom'))
'male'
>>> classifier.classify(gender_features('Kate'))
'female'
>>> accuracy(classifier, test_set)
0.796
>>> classifier.show_most_informative_features(5)
Most Informative Features
            last_letter = 'a'        female : male =     37.0 : 1.0
            last_letter = 'k'        male : female =     33.2 : 1.0
            last_letter = 'f'        male : female =     17.3 : 1.0
            last_letter = 'p'        male : female =     11.9 : 1.0
            last_letter = 'd'        male : female =      9.9 : 1.0
```

修改 gender_features()函数，为分类器提供名称的长度、它的第一个字母以及任何其他看起来可能有用的特征，再用这些新特征训练分类器并测试其准确性。结果发现准确性并没有多大提升，last_letter 才是正确的特征。

6.1.2　选择正确的特征

选择相关的特征并决定如何为一个学习方法编码它们，这对学习方法提取一个好的模型可以产生巨大的影响。建立一个分类器的很多工作之一是找出哪些特征可能是相关的以及如何能表示它们。虽然使用相当简单而明显的特征集往往可以得到像样的性能，但是使用精心构建的基于对当前任务的透彻理解的特征，通常会显著提高收益。典型地，特征提取是通过反复试验和错误建立的，有哪些信息是与问题相关的直觉指引的，包括能想到的所有特征，然后检查哪些特征是实际有用的。在例 6-1 中对名字性别特征采取这种做法。

【例 6-1】　一个特征提取器，过拟合性别特征，这个特征提取器返回的特征集包括大量指定的特征，从而导致对于相对较小的名字语料库过拟合。

```
>>> def gender_features2(name):
...     features={}
...     features["firstletter"]=name[0].lower()
...     features["lastletter"]=name[-1].lower()
...     for letter in 'abcdefghijklmnopqrstuvwxyz':
...         features["count(%s)" % letter]=name.lower().count(letter)
...         features["has(%s)" % letter]=letter in name.lower()
...     return features
...
>>> gender_features2('John')
{'count(u)': 0, 'has(d)': False, 'count(b)': 0, 'count(w)': 0,
'has(b)': False, 'count(l)': 0, 'count(q)': 0, 'count(n)': 1, 'has(j)':
True, 'count(s)': 0, 'count(h)': 1, 'has(h)': True, 'has(y)': False,
'count(j)': 1, 'has(f)': False, 'has(o)': True, 'count(x)': 0, 'has(m)':
False, 'count(z)': 0, 'has(k)': False, 'has(u)': False, 'count(d)': 0,
'has(s)': False, 'count(f)': 0, 'lastletter': 'n', 'has(q)': False,
'has(w)': False, 'has(e)': False, 'has(z)': False, 'count(t)': 0,
'count(c)': 0, 'has(c)': False, 'has(x)': False, 'count(v)': 0,
'count(m)': 0, 'has(a)': False, 'has(v)': False, 'count(p)': 0,
'count(o)': 1, 'has(i)': False, 'count(i)': 0, 'has(r)': False,
'has(g)': False, 'count(k)': 0, 'firstletter': 'j', 'count(y)': 0,
'has(n)': True, 'has(l)': False, 'count(e)': 0, 'has(t)': False,
'count(g)': 0, 'count(r)': 0, 'count(a)': 0, 'has(p)': False}
```

要用于一个给定的学习算法的特征的数目是有限的,如果提供太多的特征,那么该算法将高度依赖训练数据的特性,而一般化到新的例子的效果不会很好,这个问题被称为过拟合。运行在小训练集上时尤其会有问题,例如,如果我们使用例 6-1 中所示的特征提取器训练朴素贝叶斯分类器,将会过拟合这个相对较小的训练集,造成这个系统的精度比只考虑每个名字最后一个字母的分类器的精度低约 1%。

```
import nltk
from nltk.corpus import names
import random
names=([(name, 'male') for name in names.words('male.txt')]+
[(name, 'female') for name in names.words('female.txt')])
random.shuffle(names)
featuresets=[(gender_features2(n), g) for n, g in names]
train_set, test_set=featuresets[500:], featuresets[:500]
classifier=nltk.NaiveBayesClassifier.train(train_set)
print nltk.classify.accuracy(classifier, test_set) #0.76
```

一旦初始特征集被选定,完善特征集的一个非常有成效的方法是错误分析。首先,我们选择一个开发集,包含用于创建模型的语料数据。然后将这种开发集分为训练集和开发测试集。

```
test_names=names[:500]
devtest_names=names[500:1500]
train_names=names[1500:]
```

训练集用于训练模型,开发测试集用于进行错误分析,测试集用于系统的最终评估。将一个单独的开发测试集用于错误分析而不是使用测试集是很重要的,将语料分为适当的数据集,使用训练集训练一个模型,然后在开发测试集上运行。语料数据分为两类:开发集和测试集。

```
import nltk
from nltk.corpus import names
import random
names=([(name, 'male') for name in names.words('male.txt')]+
[(name, 'female') for name in names.words('female.txt')])
random.shuffle(names)
test_names=names[:500]
devtest_names=names[500:1500]
train_names=names[1500:]
def gender_features(word):
    return {'last_letter':word[-1]}
```

```
test_set=[(gender_features(n), g) for n, g in test_names]
devtest_set=[(gender_features(n), g) for n, g in devtest_names]
train_set=[(gender_features(n), g) for n, g in train_names]
classifier=nltk.NaiveBayesClassifier.train(train_set)
print nltk.classify.accuracy(classifier, test_set)
```

开发集通常被进一步分为训练集和开发测试集。使用开发测试集，可以生成一个分类器预测名字性别时的错误列表。

```
errors=[]
for name, tag in devtest_names:
    guess=classifier.classify(gender_features(name))
    if guess!=tag:
        errors.append((tag, guess, name))
```

然后，可以检查个别错误案例，在那里该模型预测了错误的标签，尝试确定什么额外信息将使其能够作出正确的决定(或者现有的哪部分信息导致其作出错误的决定)，然后可以相应地调整特征集。我们已经建立的名字分类器在开发测试语料上产生约100个错误。

```
for tag, guess, name in sorted(errors): # doctest: +ELLIPSIS +NORMALIZE_WHITESPACE
    print 'correct=%-8s guess=%-8s name=%-30s' % (tag, guess, name)
```

浏览这个错误列表，它明确指出一些多个字母的后缀可以指示名字性别。例如，yn结尾的名字显示以女性为主，尽管事实上，n结尾的名字往往是男性；以ch结尾的名字通常是男性，尽管以h结尾的名字倾向于是女性。因此，调整我们的特征提取器包括两个字母后缀的特征。

```
def gender_features(word):
    return {'suffix1':word[-1:], 'suffix2':word[-2:]}
```

使用新的特征提取器重建分类器，我们看到测试数据集上的性能提高了近3%。

```
def gender_features(word):
    return {'suffix1':word[-1:], 'suffix2':word[-2:]}

test_set=[(gender_features(n), g) for n, g in test_names]
devtest_set=[(gender_features(n), g) for n, g in devtest_names]
train_set=[(gender_features(n), g) for n, g in train_names]
classifier=nltk.NaiveBayesClassifier.train(train_set)
print nltk.classify.accuracy(classifier, test_set)
```

这个错误分析过程可以不断重复，检查存在于由新改进的分类器产生的错误中的模式，每一次错误分析过程被重复，我们应该选择一个不同的开发测试/训练分割，以确保该分类器不会开始反映开发测试集的特质。但是，一旦已经使用了开发测试集开发模型，无论这个模型在新数据会表现多好，我们将不能再相信它会给出一个准确的结果。因此，保持测试集分离、未使用过，直到我们的模型开发完毕是很重要的。在这一点上，可以使用测试集评估模型在新的输入值上执行得有多好。

6.1.3 文档分类

首先，构造一个标记了相应类别的文档清单。对于这个例子，我们选择 movie_reviews 语料库，将每个评论归类为 pos 或 neg。

```
>>> from nltk.corpus import movie_reviews as mr
>>> cs=mr.categories()
>>> documents=[(list(mr.words(fd)), c) for c in cs for fd in mr.fileids(c)]
>>> import random
>>> random.shuffle(documents)
>>> movie_reviews.categories() #[u'neg', u'pos']
[u'neg', u'pos']
>>> movie_reviews.fileids('neg')[:2]  #[u'neg/cv000_29416.txt', u'neg/cv001_19502.txt']
[u'neg/cv000_29416.txt', u'neg/cv001_19502.txt']
>>> movie_reviews.words('neg/cv000_29416.txt') #[u'plot', u':', u'two', u'teen', ...]
[u'plot', u':', u'two', u'teen', u'couples', u'go', ...]
```

接下来，为文档定义一个特征提取器，这样分类器就会知道哪些方面的数据应注意（见例 6-2）。对于文档主题识别，可以为每个词定义一个特性表示该文档是否包含这个词。为了限制分类器需要处理的特征的数目，我们一开始构建一个整个语料库中前 2 000 个最频繁词的列表。然后，定义一个特征提取器，简单地检查这些词是否在一个给定的文档中。

【例 6-2】 一个文档分类的特征提取器，其特征表示每个词是否在一个给定的文档中。

```
from nltk import FreqDist
all_words=FreqDist(w.lower() for w in movie_reviews.words())
word_features=all_words.keys()[:2000]
def document_features(document):
    document_words=set(document)
    features={}
    for word in word_features:
        features['contains(%s)' % word]=word in document_words
    return features
```

```
print
document_features(movie_reviews.words('pos/cv957_8737.txt'))
    document_features(movie_reviews.words('pos/cv957_8737.txt')).it
ems()[0]
    #(u'contains(corporate)', False)
```

计算文档的所有词的集合，而不仅仅检查 if word in document，因为检查一个词是否在一个集合中出现比检查它是否在一个列表中出现要快得多。

现在，已经定义了我们自己的特征提取器，可以用它来训练一个分类器，为新的电影评论加标签（例 6-3）。为了检查产生的分类器可靠性如何，在测试集上计算其准确性。可以使用 show_most_informative_features() 来找出哪些特征是分类器发现最有信息量的。

【例 6-3】 训练和测试一个分类器进行文档分类。

```
featuresets=[(document_features(d), c) for d, c in documents]
train_set, test_set=featuresets[100:], featuresets[:100]
classifier=nltk.NaiveBayesClassifier.train(train_set)
print nltk.classify.accuracy(classifier, test_set) #0.75
classifier.show_most_informative_features(5)
```

显然在这个语料库中，提到 Seagal 的评论中负面数据条数约是正面数据条数的 8 倍，而提到 Damon 的评论中正面数据条数约是负面数据条数的 6 倍。

6.1.4 词性标注

可以训练一个分类器来算出哪个后缀最有信息量。首先，找出最常见的后缀。

```
Python 2.7.11 |Anaconda 2.5.0 (64-bit)| (default, Jan 29 2016,
14:26:21) [MSC v.1500 64 bit (AMD64)] on win32
>>> from nltk.corpus import brown
>>> from nltk import FreqDist
>>> suffix_fdist=FreqDist()
>>> for word in brown.words():
...     word=word.lower()
...     suffix_fdist[word[-1:]]+=1
...     suffix_fdist[word[-2:]]+=1
...     suffix_fdist[word[-3:]]+=1
...
>>> common_suffixes=suffix_fdist.most_common()[:10]
>>> print common_suffixes
[(u'e', 202946), (u',', 175002), (u'.', 152999), (u's', 128722),
```

(u'd', 105687), (u't', 94459), (u'he', 92084), (u'n', 87889), (u'a', 74912), (u'of', 72978)]

接下来，定义一个特征提取器函数，检查给定单词的这些后缀。

```
def pos_features(word):
    features={}
    for suffix in common_suffixes:
        features['endswith(%s)' % suffix]=word.lower().endswith(suffix)
    return features
```

特征提取函数的行为就像有色眼镜一样，强调数据中的某些属性（颜色），并使其无法看到其他属性。分类器在决定如何标记输入时，将完全依赖它们强调的属性。在这种情况下，分类器将只基于一个给定的词拥有（如果有）哪个常见后缀的信息来做决定。

现在，已经定义了我们自己的特征提取器，可以用它来训练一个新的"决策树"分类器。

```
tagged_words=brown.tagged_words(categories='news')[:1000]
featuresets=[(pos_features(n), g) for n, g in tagged_words]
size=int(len(featuresets)*0.1)
train_set, test_set=featuresets[size:], featuresets[:size]
classifier=nltk.DecisionTreeClassifier.train(train_set)
nltk.classify.accuracy(classifier, test_set) #0.65
classifier.classify(pos_features('cats')) #'NNS'
```

决策树模型的一个很好的性质是它们往往很容易解释。甚至可以指示 NLTK 将它们以伪代码形式输出。

```
print classifier.pseudocode(depth=4)
```

在这里，可以看到分类器一开始检查一个词是否以逗号结尾。如果是，它会得到一个特别的标记","。接下来，分类器检查词是否以"the"结尾，这种情况它几乎肯定是一个限定词。这个"后缀"被决策树早早使用是因为词 the 太常见。分类器继续检查词是否以 s 结尾，如果是，那么它极有可能得到动词标记 VBZ（says，除非它是这个词 is，有特殊标记 BEZ）；如果不是，那么它往往是名词（除非它是标点符号"."）。实际的分类器包含这里显示的 if-then 语句下面进一步的嵌套，参数 depth=4 只显示决定树的顶端部分。

6.1.5 探索上下文语境

只要特征提取器仅仅看着目标词，我们就没法添加目标词出现的上下文语境特征。然而，上下文语境特征往往提供关于正确标记的强大线索，例如，标注词 fly，如果知道它前面的词是 a，将使我们能够确定它是一个名词，而不是一个动词。为了采取基于词的上下文的特征，我们必须修改以前为特征提取器定义的模式，不是只传递未标注的词，将传递整个未标注的句子以及目标词的索引。例 6-4 演示了这种方法，使用依赖上下文特征的提取器定义一个词性分类器。

【例 6-4】 一个词性分类器，它的特征提取器检查一个词出现的上下文，以便决定应该分配的词性标记，特别地，前面的词作为一个特征。

```
>>> from nltk.corpus import brown
>>> from nltk.tag import untag
>>> from nltk import NaiveBayesClassifier
>>> from nltk.classify import accuracy
>>> def pos_features(sentence, i):
...     features={"suffix(1)": sentence[i][-1:], "suffix(2)": sentence[i][-2:],
...  "suffix(3)": sentence[i][-3:]}
...     if i==0:
...         features["prev-word"]="<START>"
...     else:
...         features["prev-word"]=sentence[i-1]
...     return features
...
>>> tagged_sents=brown.tagged_sents(categories='news')
>>> featuresets=[]
>>> for tagged_sent in tagged_sents:
...     untagged_sent=untag(tagged_sent)
...     for i, (word, tag) in enumerate(tagged_sent):
...         featuresets.append((pos_features(untagged_sent, i), tag))
...
>>> size=int(len(featuresets)*0.1)
>>> train_set, test_set=featuresets[size:], featuresets[:size]
>>> classifier=NaiveBayesClassifier.train(train_set)
>>> accuracy(classifier, test_set) #0.7891596220785678
0.7891596220785678
```

很显然，利用上下文特征提高了词性标注器的性能，例如，分类器学到一个词如果紧跟在词 large 或 gubernatorial 后面，极可能是名词。然而，它无法学到一个词如果它跟在形容词后面可能是名词，这样更一般，因为它没有获得前面这个词的词性标记。

6.1.6 序列分类

一种序列分类策略，称为连续分类或贪婪序列分类，是为第一个输入找到最有可能的类标签，然后使用这个问题的答案帮助找到下一个输入的最佳标签，这个过程可以不断重复直到所有的输入都被贴上标签。在例 6-5 演示了这一策略，首先，扩展特征提取函数使其具有参数 history，它提供一个到目前为止已经为句子预测标记的列表。注意，在训练时，使用已标注的标记为特征提取器提供适当的历史信息，但标注新句子 tag 时，基于标注器本身的输出产生历史信息，history 中的每个标记对应句子中的一个词。注意，history 只包含已经归类词的标记，也就是目标词左侧的词，因此，虽然是有可能查看目标词右边词的某些特征，但查看那些词的标记是不可能的，因为还未产生它们。已经定义了特征提取器，便可以继续建立序列分类器。

【例 6-5】 使用连续分类器进行词性标注。

```
import nltk
from nltk import TaggerI
from nltk.tag import untag

from nltk.corpus import brown
def pos_features(sentence, i, history):
    features={"suffix(1)": sentence[i][-1:], "suffix(2)": sentence[i][-2:], "suffix(3)": sentence[i][-3:]}
    if i==0:
        features["prev-word"]="<START>"
        features["prev-tag"]="<START>"
    else:
        features["prev-word"]=sentence[i-1]
        features["prev-tag"]=history[i-1]
    return features
#tagger=ConsecutivePosTagger(train_sents)

class ConsecutivePosTagger(TaggerI):
    def __init__(self, train_sents):
        train_set=[]
        for tagged_sent in train_sents:
            untagged_sent=untag(tagged_sent)
            history=[]
            for i, (word, tag) in enumerate(tagged_sent):
                featureset=pos_features(untagged_sent, i, history)
                train_set.append((featureset, tag))
                history.append(tag)
        self.classifier=nltk.NaiveBayesClassifier.train(train_set)
    def tag(self, sentence):
        history=[]
```

```
            for i, word in enumerate(sentence):
                featureset=pos_features(sentence , i, history)
                tag=self.classifier.classify(featureset)
                history.append(tag)
            return zip(sentence, history)

tagged_sents=brown.tagged_sents(categories='news')
size=int(len(tagged_sents)*0.1)
train_sents, test_sents=tagged_sents[size:], tagged_sents[:size]
tagger=ConsecutivePosTagger(train_sents)
print tagger.evaluate(test_sents) #0.79796012981
tagger.tag(['These', 'are', 'cats'])


>>> import nltk
>>> from nltk import TaggerI
>>> from nltk.tag import untag
>>>
>>> from nltk.corpus import brown
>>> def pos_features(sentence, i, history):
...     features={"suffix(1)": sentence[i][-1:], "suffix(2)": sentence[i][-2:], "
suffix(3)": sentence[i][-3:]}
...     if i==0:
...         features["prev-word"]="<START>"
...         features["prev-tag"]="<START>"
...     else:
...         features["prev-word"]=sentence[i-1]
...         features["prev-tag"]=history[i-1]
...     return features
... #tagger=ConsecutivePosTagger(train_sents)
...
>>> class ConsecutivePosTagger(TaggerI):
...     def __init__(self, train_sents):
...         train_set=[]
...         for tagged_sent in train_sents:
...             untagged_sent=untag(tagged_sent)
...             history=[]
...             for i, (word, tag) in enumerate(tagged_sent):
...                 featureset=pos_features(untagged_sent, i, history)
```

```
...             train_set.append((featureset, tag))
...             history.append(tag)
...         self.classifier=nltk.NaiveBayesClassifier.train(train_set)
...   def tag(self, sentence):
...         history=[]
...         for i, word in enumerate(sentence):
...             featureset=pos_features(sentence , i, history)
...             tag=self.classifier.classify(featureset)
...             history.append(tag)
...         return zip(sentence, history)
...
>>> tagged_sents=brown.tagged_sents(categories='news')
>>> size=int(len(tagged_sents)*0.1)
>>> train_sents, test_sents=tagged_sents[size:], tagged_sents[:size]
>>> tagger=ConsecutivePosTagger(train_sents)
>>> print tagger.evaluate(test_sents)  #0.79796012981
0.798052851182
>>> tagger.tag(['These', 'are', 'cats'])
[('These', u'DTS'), ('are', u'BER'), ('cats', u'NNS')]
```

6.2 有监督分类的典型应用

6.2.1 句子分割

句子分割可以看作是一个标点符号的分类，每当我们遇到一个可能会结束一个句子的符号，如句号或问号，我们必须决定它是否终止了当前句子。第一步是获得一些已被分割成句子的数据，将它转换成一种适合提取特征的形式。

```
sents=[['This', 'is', 'a', 'book', '.'], ['That', 'is', 'a', 'desk', '.']]
tokens=[]
boundaries=set()
offset=0
for sent in sents:
...     tokens.extend(sent)
...     offset+=len(sent)
...     boundaries.add(offset-1)
...
tokens
```

```
['This', 'is', 'a', 'book', '.', 'That', 'is', 'a', 'desk', '.']
boundaries
{9, 4}
```

This	is	a	book	.	That	is	a	desk	.
0	1	2	3	4	5	6	7	8	9

在这里，tokens 是所有句子单词合并的列表，boundaries 是一个包含所有句子最后一个单词索引的集合。下一步，需要指定用于决定标点是否表示句子边界的数据特征。

```
def punct_features(tens, i):
    return {'next-word-capitalized':tens[i+1][0].isupper(),
    'prevword':tens[i-1].lower(),
    'punct':tens[i],
    'prev-word-is-one-char':len(tens[i-1])==1}
```

基于这一特征提取器，可以通过选择所有的标点符号创建一个加标签的特征集的列表，然后标注它们是否是边界标识符。

```
featuresets=[(punct_features(tens, i), (i in boundaries))
    for i in range(1, len(tens)-1) if tens[i] in '.?!']
```

使用这些特征集可以训练和评估一个标点符号分类器。

```
size=int(len(featuresets)*0.1)
train_set, test_set=featuresets[size:], featuresets[:size]
classifier=nltk.NaiveBayesClassifier.train(train_set)
nltk.classify.accuracy(classifier, test_set) #0.9274193548387096
```

使用这种分类器进行断句，只需检查每个标点符号，看它是否是作为一个边界标识符，在边界标识符处分割词列表。例 6.8 中的清单显示了如何做到这一点。

【例 6-6】 基于分类的断句器。

```
def segment_sentences(words):
    start=0
    sents=[]
    for i, word in enumerate(words):
        if word in '.?!' and classifier.classify(punct_features(words, i)) == True:
            sents.append(words[start:i+1])
            start=i+1
    if start<len(words):
        sents.append(words[start:])
    print sents
```

```
words=['Excuse', 'me', '.',
'Is', 'there', 'any', 'summer', 'resort', 'around', 'the', 'town', '?']
segment_sentences(words[:-1])
```

6.2.2 识别对话行为类型

处理对话时,将对话看作说话者执行的动作是很有用的,对于表述行为的陈述句这种解释是最简单的,但是问候、问题、回答、断言和说明都可以是基于语言的行动类型。识别对话中言语的对话行为是理解谈话的第一步。NPS 聊天语料库包括超过 10 000 个来自即时消息会话的帖子,这些帖子都已经被贴上 15 种对话行为类型中的一种标签。

第一步,提取基本的消息数据,调用 xml_posts()得到一个数据结构,表示每个帖子的 XML 注释。

```
>>> from nltk import NaiveBayesClassifier, word_tokenize
>>> from nltk.classify import accuracy
>>> from nltk.corpus import nps_chat
>>> posts=nps_chat.xml_posts()[:10]
>>> posts
[<Element 'Post' at 0x00000000062E40F0>, <Element 'Post' at 0x00000000062FA270>, ...]
```

第二步,定义一个简单的特征提取器,检查帖子包含什么词。

```
def dialogue_act_features(post):
    features={}
    for word in word_tokenize(post):
        features['contains(%s)' % word.lower()]=True
    return features
```

第三步,通过为每个帖子提取特征(使用 post.get('class')获得一个帖子的对话行为类型)构造训练和测试数据,并创建一个新的分类器。

```
featuresets=[(dialogue_act_features(post.text),
post.get('class')) for post in posts]
    size=int(len(featuresets)*0.1)
    train_set, test_set=featuresets[size:], featuresets[:size]
    classifier=NaiveBayesClassifier.train(train_set)
    print accuracy(classifier, test_set)
    classifier.classify(dialogue_act_features("I forgive you ."))
#'Statement'
```

6.2.3 识别文字蕴含

识别文字蕴含（RTE）是判断文本 T 的一个给定片段是否蕴含着另一个叫作"假设"的文本。迄今为止，已经有 4 个 RTE 挑战赛，在那里共享的开发和测试数据会提供给参赛队伍，这里是挑战赛 3 开发数据集中的文本/假设对的例子。标签 True 表示蕴含成立，False 表示蕴含不成立。

应当强调，文字和假设之间的关系并不一定是逻辑蕴涵，而是一个人是否会得出结论。文本提供了合理的证据证明假设是真实的。

可以把 RTE 当作一个分类任务，尝试为每一对预测真/假标签，这项任务的成功做法看上去涉及语法分析、语义和现实世界知识的组合。RTE 许多早期的尝试使用粗浅的分析基于文字和假设之间在词级别的相似性取得了相当不错的结果，在理想情况下，我们希望如果有一个蕴涵，那么假设所表示的所有信息也应该在文本中表示，相反，如果假设中有的资料文本中没有，那么就没有蕴涵。

在 RTE 特征探测器中（见例 6-7），让词（即词类型）作为信息的代理，特征计数词重叠的程度和假设中有而文本中没有词的程度（由 hyp_extra()方法获取）。不是所有的词都是同样重要的，命名实体，如人、组织和地方的名称，可能会更为重要，这促使我们分别为 words 和 nes（命名实体）提取不同的信息。此外，一些高频虚词作为停用词会被过滤掉。

【例 6-7】识别文字蕴含的特征提取器，RTEFeatureExtractor 类建立了一个除去一些停用词后在文本和假设中都有的词汇包，然后计算重叠和差异。

```
from nltk import RTEFeatureExtractor
def rte_features(rtepair):
    extractor=RTEFeatureExtractor(rtepair)
    features={}
    features['word_overlap']=len(extractor.overlap('word'))
    features['word_hyp_extra']=len(extractor.hyp_extra('word'))
    features['ne_overlap']=len(extractor.overlap('ne'))
    features['ne_hyp_extra']=len(extractor.hyp_extra('ne'))
    return features
```

为了说明这些特征的内容，我们检查前面显示的文本/假设对的一些属性。

```
from nltk.corpus import rte
from nltk import RTEFeatureExtractor
rtepair=rte.pairs(['rte3_dev.xml'])[33]
extractor=RTEFeatureExtractor(rtepair)
print extractor.text_words
print extractor.hyp_words #set(['member', 'SCO', 'China'])
print extractor.overlap('word') #set([])
print extractor.overlap('ne') #set(['SCO', 'China'])
print extractor.hyp_extra('word') #set(['member'])
```

这些特征表明假设中所有重要的词都包含在文本中，因此有一些证据支持标记这个为 True。nltk.classify.rte_classify 模块使用这些方法在合并的 RTE 测试数据上取得了超过 58%的准确率。

6.3 评　估

为了决定一个分类模型是否准确地捕捉了模式，我们必须评估该模型，评估的结果对于决定模型多么值得信赖以及我们如何使用它是非常重要的。评估是一个有效的工具，用于指导我们在未来改进模型。

6.3.1 测试集

大多数评估技术为模型计算一个得分，通过比较它在测试集或评估集中为输入生成的标签与那些输入的正确标签，该测试集通常与训练集具有相同的格式。然而，测试集与训练语料不同是非常重要的，如果简单地重复使用训练集作为测试集，那么一个只记住了它的输入而没有学会如何推广到新例子的模型会得到误导人的高分。

建立测试集时，往往是一个可用于测试的和可用于训练的数据量之间的权衡，对于有少量平衡的标签和一个多样化测试集的分类任务，只要 100 个评估实例就可以进行有意义的评估，但是，如果一个分类任务有大量的标签或包括罕见的标签，那么选择的测试集的大小就要保证出现次数最少的标签至少出现 50 次。此外，如果测试集包含许多密切相关的实例，例如，来自一个单独文档中的实例，那么测试集的大小应增加，以确保这种多样性的缺乏不会扭曲评估结果。当有大量已标注数据可用时，只使用整体数据的 10%进行评估常常会在安全方面犯错。选择测试集时另一个需要考虑的是测试集中实例与开发集中的实例的相似程度，这两个数据集越相似，我们对评估结果推广到其他数据集的信心就越小，例如，考虑词性标注任务，在一种极端情况，可以通过从一个反映单一文体的数据源随机分配句子，创建训练集和测试集。

```
from random import shuffle
from nltk.corpus import brown
tagged_sents=list(brown.tagged_sents(categories='news'))
shuffle(tagged_sents)
size=int(len(tagged_sents)*0.1)
train_set, test_set=tagged_sents[size:], tagged_sents[:size]
```

在这种情况下，测试集和训练集将是非常相似的，训练集和测试集均取自同一文体，所以不能相信评估结果可以推广到其他文体。一个稍好的做法是确保训练集和测试集来自不同的文件。

```
from nltk.corpus import brown
file_ids=brown.fileids(categories='news')
size=int(len(file_ids)*0.1)
train_set=brown.tagged_sents(file_ids[size:])
test_set=brown.tagged_sents(file_ids[:size])
```

如果要执行更令人信服的评估,可以从与训练集中文档联系更少的文档中获取测试集。

```
from nltk.corpus import brown
train_set=brown.tagged_sents(categories='news')
test_set=brown.tagged_sents(categories='fiction')
```

如果在此测试集上建立了一个性能很好的分类器,那么完全可以相信它有能力很好地泛化到用于训练它的数据以外的数据。试计算这三种测试集的得分。

```
from random import shuffle
from nltk.corpus import brown
tagged_sents=list(brown.tagged_sents(categories='news'))
shuffle(tagged_sents)
size=int(len(tagged_sents)*0.1)
train_set, test_set=tagged_sents[size:], tagged_sents[:size]
from nltk.corpus import brown
file_ids=brown.fileids(categories='news')
size=int(len(file_ids)*0.1)
train_set=brown.tagged_sents(file_ids[size:])
test_set=brown.tagged_sents(file_ids[:size])
from nltk.corpus import brown
train_set=brown.tagged_sents(categories='news')
test_set=brown.tagged_sents(categories='fiction')
```

6.3.2 准确度

用于评估一个分类器最简单的是准确度。测量测试集上分类器正确标注输入的比例,例如,一个名字性别分类器,在包含 80 个不同名字的测试集上预测正确的名字有 60 个,它有 60/80=75%的准确度。nltk.classify.accuracy()函数会在给定的测试集上计算分类器模型的准确度。

```
>>> def gender_features(w):
...     return {'last_letter':w[-1]}
...
>>> from nltk.corpus import names
>>> from random import shuffle
```

```
>>> names=[(n, 'male') for n in names.words('male.txt')]+[(n,
'female') for n in names.words('female.txt')]
>>> shuffle(names)
>>> featuresets=[(gender_features(n), g) for n, g in names]
>>> train_set, test_set=featuresets[500:], featuresets[:500]
>>> from nltk import NaiveBayesClassifier
>>> classifier=NaiveBayesClassifier.train(train_set)
>>> print 'Accuracy: %4.2f' % accuracy(classifier, test_set)
Accuracy: 0.77
```

解释一个分类器的准确性得分，考虑测试集中单个类标签的频率是很重要的，例如，考虑一个决定词 bank 每次出现正确词义的分类器，如果我们在金融新闻文本上评估分类器，那么可能会发现，金融机构的意思 20 个里面出现了 19 次，在这种情况下，19/20=1-1/20=1-0.05=0.95=95%的准确度也是有问题的，因为我们可以构建一个模型，它总是返回金融机构的意义。换句话说，准确度有局限性，然而，如果我们在一个更加平衡的语料库上评估分类器，那里最频繁的词义只占 40%，那么 95%的准确度得分将是一个更加有益的结果。

6.3.3 精确度和召回率

准确度分数可能会产生误导，例如在搜索任务中，如信息检索，我们试图找出与特定任务有关的文档，由于不相关文档的数量远远多于相关文档的数量，一个将每一个文档都标记为无关的模型准确度分数将非常接近 100%，即使将相关文档都准确标记为相关的模型准确度分数将非常接近 0%。这种情况用准确度分数来评价就不科学了，因此，对搜索任务使用不同的测量集是很常见的。

真阳性（TP）是相关项目中我们正确识别为相关的。真阴性（TF）是不相关项目中我们正确识别为不相关的。假阳性（FP）或Ⅰ型错误是不相关项目中我们错误识别为相关的。假阴性（FN）或Ⅱ型错误是相关项目中我们错误识别为不相关的。给定这四个数字，我们可以定义以下指标。

精确度（Precision）：表示我们发现的项目中有多少是相关的，TP/(TP+FP)。

召回率（Recall）：表示相关的项目中我们发现了多少，TP/(TP+FN)。

F-度量值(F-Measure)或 F-得分（F-Score），组合精确度和召回率为一个单独的得分，被定义为精确度和召回率的调和平均数 2×Precision×Recall/(Precision+Recall)。

6.3.4 混淆矩阵

一个混淆矩阵是一个表，其中每个 cells[i, j]表示正确的标签 i 被预测为标签 j 的次数。因此，对角线项目，即 cells[i, i]表示正确预测的标签，非对角线项目表示错误。

```
Deactivating environment "C:\Anaconda2"...
Activating environment "C:\Anaconda2"...
```

```
[Anaconda2] C:\Users\PC>python
Python 2.7.11 |Anaconda 2.5.0 (64-bit)| (default, Jan 29 2016, 14:26:21) [MSC v.1500 64 bit (AMD64)] on win32
Type "help", "copyright", "credits" or "license" for more information.
Anaconda is brought to you by Continuum Analytics.
Please check out: http://continuum.io/thanks and https://anaconda.org
>>> from nltk.corpus import brown
>>> from nltk import DefaultTagger, UnigramTagger, ConfusionMatrix
>>> from nltk.tag import untag
>>> brown_tagged_sents=brown.tagged_sents(categories='news')
>>> brown_sents=brown.sents(categories='news')
>>> size=int(len(brown_tagged_sents)*0.9)
>>> train_sents=brown_tagged_sents[:size]
>>> test_sents=brown_tagged_sents[size:]
>>> t0=DefaultTagger('NN')
>>> t2=UnigramTagger(train_sents, backoff=t0)
>>> def tag_list(tagged_sents):
...     return [tag for sent in tagged_sents for word, tag in sent]
...
>>> def apply_tagger(tagger, corpus):
...     return [tagger.tag(untag(sent)) for sent in corpus]
...
>>> gold=tag_list(brown.tagged_sents(categories='editorial')[:7])
>>> test=tag_list(apply_tagger(t2, brown.tagged_sents(categories='editorial'))[:7])
>>> cm=ConfusionMatrix(gold, test)
>>> print cm
    |                                       V            |
    |     A           J    N N              B            |
    |     P           B    J N N            P    D       |
    |  -  E  B  B  -  -  N  P  P  P  V  -  V  V  W  |
    |     A  H  A  D  E  E  C  C  C  D  E  H  V  I  J  T  M  N  H  |
    |     T  N  N  P  P  S  R  T  V  B  H  B  B  B  D     |
    |  *  ,  .  P  L  T  Z  N  Z  C  D  S  T  X  V  Z  N  J  L  D  N  |
    |     L  L  S  R  O  S  S  B  O  B  D  L  G  N  Z  T     |
----+----------------------------------------------------+
  * | <1> . . . . . . . . . . . . . . . . . . . . . . . |
```

```
      ,   | . .<6>. . . . . . . . . . . . . . . . . . . . . . . . . . . . . . . . . . . . . |
      .   | . .<6>. . . . . . . . . . . . . . . . . . . . . . . . . . . . . . . . . . . . . |
     AP   | . . .<2>. . . . . . . . . . . . . . . . . . . . . . . . . . . . . . . . . . . . |
   AP-HL  | . . . 1 <.>. . . . . . . . . . . . . . . . . . . . . . . . . . . . . . . . . . . |
     AT   | . . . . .<16>. . . . . . . . . . . . . . . . . . . . . . . . . . . . . . . . . . |
    BEDZ  | . . . . . .<2>. . . . . . . . . . . . . . . . . . . . . . . . . . . . . . . . . |
     BEN  | . . . . . . .<1>. . . . . . . . . . . . . . . . . . . . . . . . . . . . . . . . |
     BEZ  | . . . . . . . .<1>. . . . . . . . . . . . . . . . . . . . . . . . . . . . . . . |
     CC   | . . . . . . . . .<5>. . . . . . . . . . . . . . . . . . . . . . . . . . . . . . |
     CD   | . . . . . . . . . .<1>. . . . . . . . . . . . . . . . . . . . . . . . . . . . . |
     CS   | . . . . . . . . . . .<1>. . . . . . . . . . . . . . . . . . . . . . . . . . . . |
     DT   | . . . . . . . . . . . .<1>. . . . . . . . . . . . . . . . . . . . . . . . . . . |
     EX   | . . . . . . . . . . . . .<1>. . . . . . . . . . . . . . . . . . . . . . . . . . |
     HV   | . . . . . . . . . . . . . .<2>. . . . . . . . . . . . . . . . . . . . . . . . . |
    HVZ   | . . . . . . . . . . . . . . .<2>. . . . . . . . . . . . . . . . . . . . . . . . |
     IN   | . . . . . . . . . . . . . . . .<14>. . . . . . . . 1 . . . . . . . . . . . . . . |
     JJ   | . . . . . . . . . . . . . . . . .<3>. . 1 . . . . . . . . . . . . . . . . . . . |
    JJ-TL | . . . . . . . . . . . . . . . . . . .<2>. . . . . . . . . . . . . . . . . . . . |
     MD   | . . . . . . . . . . . . . . . . . . . .<1>. . . . . . . . . . . . . . . . . . . |
     NN   | . . . . . . . . . . . . . . . . . 1 . .<23>. . . . . . . . . . . . . . . 1 . . . |
    NN-HL | . . . . . . . . . . . . . . . . . 1 . . 1 <.> 1 . . . . . . . . . . . . . . . . . |
    NN-TL | . . . . . . . . . . . . . . . . . . . . . .<2>. . . . . . . . . . . . . . . . . |
     NNS  | . . . . . . . . . . . . . . . . . . . . 1 . .<8>. . . . . . . . . . . . . 1 . . |
     NR   | . . . . . . . . . . . . . . . . . . . . . . . .<1>. . . . . . . . . . . . . . . |
     PPO  | . . . . . . . . . . . . . . . . . . . . . . . . .<1>. . . . . . . . . . . . . . |
     PPS  | . . . . . . . . . . . . . . . . . . . . . . . . . .<2>. . . . . . . . . . . . . |
    PPSS  | . . . . . . . . . . . . . . . . . . . . . . . . . . .<1>. . . . . . . . . . . . |
     RB   | . . . . . . . . . . . . . . . . . . . . . . . . . . . .<3>. . . . . . . . . . . |
     TO   | . . . . . . . . . . . . . . . . . . . . . . . . . . . . .<5>. . . . . . . . . . |
     VB   | . . . . . . . . . . . . . . . . . . 3 . . . . . . . . . . .<3>. . . . . . . . . |
     VBD  | . . . . . . . . . . . . . . . . . . . 1 . . . . . . . . . . .<2>. . 1 . . . . . |
   VBD-HL | . . . . . . . . . . . . . . . . . . . . . . . . . . . . . . . . 1 <.>. . . . . . |
     VBG  | . . . . . . . . . . . . . . . . . . 3 . . . . . . . . . . . . . .<.>. . . . . . |
     VBN  | . . . . . . . . . . . . . . . . . . . 1 . . . . . . . . . . . 1 . .<4>. . . . . |
     VBZ  | . . . . . . . . . . . . . . . . . . . 1 . . . . . . . . . . . . . . .<.>. . . . |
     WDT  | . . . . . . . . . . . . . . . . . . . . . . . . . . . . . . . . . . . . . .<3>|
   -------+---------------------------------------------------------------------------------+
(row = reference; col = test)
```

这个混淆矩阵显示常见的错误，包括 NN 替换成了 JJ1.6%的词，NN 替换为了 NNS1.5%的词。注意：点（.）表示值为 0 的单元格，对角线项目，对应正确的分类，用尖括号标记。

6.4 决策树

你是否玩过"二十个问题"的游戏，游戏的规则很简单，参与游戏的一方在脑海里想某个事物，其他参与者向他提问题，只允许提 20 个问题，问题的答案也只能用对或错回答。问问题的人通过推断分解，逐步缩小待猜测事物的范围。决策树的工作原理与 20 个问题类似，用户输入一系列数据，然后给出游戏的答案。

6.4.1 决策树的构造

在构造决策树时，我们需要解决的第一个问题是，当前数据集上哪个特征在划分数据分类时起决定性作用。为了找到决定性的特征，划分出最好的结果，我们必须评估每个特征。完成测试之后，原始数据集就被划分为几个数据子集。这些数据子集会分布在第一个决策点的所有分支上。如果某个分支下的数据属于同一类型，则当前无须阅读的垃圾邮件已经正确地划分数据分类，无须进一步对数据集进行分割。如果数据子集内的数据不属于同一类型，则需要重复划分数据子集的过程。如何划分数据子集的算法和划分原始数据集的方法相同，直到所有具有相同类型的数据均在一个数据子集内。

表 6-1 的数据包含 5 个海洋动物，特征包括不浮出水面、是否可以生存、是否有脚蹼。我们可以将这些动物分成两类：鱼类和非鱼类。现在想要决定依据第一个特征还是第二个特征划分数据。在回答这个问题之前，必须采用量化的方法判断如何划分数据。

表 6-1 海洋生物数据

不浮出水面	是否可以生存	是否有脚蹼	属于鱼类
1	是	是	是
2	是	是	是
3	是	否	否
4	否	是	否
5	否	是	否

6.4.2 熵和信息增益

当我们用给定的特征分割输入值时，信息增益衡量它们变得更有序的程度，要衡量原始输入值集合如何无序，可以计算它们标签的熵。如果输入值的标签非常不同，熵就高；如果输入值的标签都相同，熵就低。熵被定义为每个标签的概率乘以那个标签的 log 概率的总和。

$$H = -\sum_{l \in \text{labels}} P(l) \log_2 P(l)$$

注意，如果大多数输入值具有相同的标签，例如，如果 P(male)接近 0 或接近 1，那么熵很低。特别地，低频率的标签不会贡献多少给熵，因为 P(l)很小，高频率的标签对熵也没有多大帮助，因为 $\log_2 P(l)$很小。如果输入值的标签变化很多，那么有很多中等频率的标签，它们的 P(l)和 $\log_2 P(l)$都不小，所以熵很高。

【例 6-8】 计算标签列表的熵。

```
>>> from nltk import FreqDist
>>> from math import log
>>> def entropy(labels):
...     freqdist=FreqDist(labels)
...     probs=[freqdist.freq(l) for l in freqdist]
...     return -sum(p*log(p, 2) for p in probs)
...
>>> print entropy(['male', 'male', 'male', 'male']) #0.0
-0.0
>>> print entropy(['male', 'female', 'male', 'male']) #0.811278124459
0.811278124459
>>> print entropy(['female', 'male', 'female', 'male']) #1.0
1.0
>>> print entropy(['female', 'female', 'male', 'female']) #0.811278124459
0.811278124459
>>> print entropy(['female', 'female', 'female', 'female']) #0.0
-0.0
labels=['male', 'male', 'male', 'male']
freqdist=FreqDist(labels)
freqdist
freqdist.freq('male')
labels=['male', 'female', 'male', 'male']
freqdist=FreqDist(labels)
freqdist.freq('male')
freqdist.freq('female')
```

一旦已经计算原始输入值的标签集的熵，就可以判断应用了决策树桩之后标签会变得多么有序。为了这样做，我们计算每个决策树桩的叶子的熵，利用这些叶子熵值的平均值（加权每片叶子的样本数量）。信息增益等于原来的熵减去这个新的减少的熵。信息增益越高，将输入值分为相关组的决策树桩就越好，于是可以通过选择具有最高信息增益的决策树桩来建立决策树。

决策树的另一个考虑因素是效率。前面描述的选择决策树桩的简单算法必须为每一个可能的特征构建候选决策树桩，并且这个过程必须在构造决策树的每个节点上不断重复。已经开发了一些算法通过存储和重用先前评估的例子的信息减少训练时间。

决策树有一些有用的性质。首先，它们简单明了、容易理解，决策树顶部附近尤其如此，这通常使学习算法可以找到非常有用的特征。决策树特别适合有很多层次的分类区别的情况，例如，决策树可以非常有效地捕捉进化树。

然而，决策树也有一些缺点。第一个问题是，由于决策树的每个分支会划分训练数据，在训练树的低节点，可用的训练数据量可能会变得非常小。因此，这些较低的决策节点可能过拟合训练集，学习模式反映训练集的特质而不是问题底层显著的语言学模式。对这个问题的一个解决方案是，训练数据量变得太小时停止分裂节点。另一种方案是长出一个完整的决策树，但随后进行剪枝，剪去在开发测试集上不能提高性能的决策节点。

决策树的第二个问题是，它们强迫特征按照一个特定的顺序进行检查，即使特征可能是相对独立的。例如，按主题分类文档（如体育、汽车等）时，特征如 hasword(football)，极可能表示一个特定标签，无论其他的特征值是什么。由于决定树顶部附近的空间有限，大部分这些特征将需要在树中的许多不同的分支中重复。因为越往树的下方，分支的数量呈指数倍增长，重复量可能非常大。

一个相关的问题是决策树不善于利用对正确的标签具有较弱预测能力的特征。由于这些特征的影响相对较小，它们往往出现在决策树非常低的地方。决策树学习的时间远远不够用到这些特征，也不能留下足够的训练数据来可靠地确定它们应该有什么样的影响。如果我们能够在整个训练集中看看这些特征的影响，那么也许能够做出一些关于它们是如何影响标签选择的结论。

决策树需要按一个特定的顺序检查特征的事实，限制了它们利用相对独立的特征的能力。下面将讨论的朴素贝叶斯分类方法克服了这一限制，允许所有特征"并行"地起作用。

6.5 朴素贝叶斯分类器

6.5.1 核心思想

对于给出的待分类项，求解在此项出现的条件下各个类别出现的概率，哪个最大，就认为此待分类项属于哪个类别。

6.5.2 概率基础

1. 随机变量的先验、条件、联合概率

先验概率：事件发生前的预判概率 $P(X)$。

条件概率：一个事件发生后另一个事件发生的概率 $P(X_1|X_2)$，$P(X_2|X_1)$。

联合概率：是指两个事件同时发生的概率 $X=(X_1, X_2)$，$P(X)=P(X_1, X_2)$。

相关：$P(X_1, X_2)=P(X_2|X_1)P(X_1)=P(X_1|X_2)P(X_2)$。

独立：$P(X_1, X_2)=P(X_1)P(X_2)$。

2. 朴素贝叶斯的公式及其作用

贝叶斯公式打通从 $P(A|B)$ 获得 $P(B|A)$ 的通路。

$$P(A|B) = P(A)\frac{P(B|A)}{P(B)}$$

式中，$P(A|B), P(B|A)$ 表示后验概率；$P(A)$ 表示先验概率；$\frac{P(B|A)}{P(B)}$ 表示可能性函数。

3. 条件概率和联合概率的计算

将一枚硬币抛掷，观察其出现正反面的情况。假设事件 A 为"两次掷出同一面"，事件 B 为"至少有一次为正面"，现在来求已知事件 B 已经发生的条件下事件 A 发生的概率。

样本空间{HH，HT，TH，TT}

样本 A={HH，TT}

样本 B={HH，HT，TH}

$$P(A) = \frac{2}{4} = \frac{1}{2}$$

$$P(B) = \frac{3}{4}$$

$$P(AB) = \frac{1}{4} = P(A|B)P(B) = \frac{1}{3} \times \frac{3}{4}$$

$$P(A|B) = \frac{1}{3} = \frac{P(AB)}{P(B)} = \frac{1/4}{3/4}$$

6.5.3 朴素贝叶斯分类器的应用

一所学校里面有 60%的男生，40%的女生，男生总是穿裤子，女生则一半穿裤子一半穿裙子。假设你走在校园中，前面走着一个穿裤子的学生，你能够推断出这个学生是女生的概率是多大？现在假设学校里面人的总数为 100。

（1）男生中穿裤子的人数：$100 \times P(\text{Boy}) \times P(\text{Pants}|\text{Boy}) = 100 \times 60\% \times 100\% = 60$；

（2）女生中穿裤子的人数：$100 \times P(\text{Girl}) \times P(\text{Pants}|\text{Girl}) = 100 \times 40\% \times 50\% = 20$；

（3）穿裤子的总人数：$100 \times P(\text{Boy}) \times P(\text{Pants}|\text{Boy}) + 100 \times P(\text{Girl}) \times P(\text{Pants}|\text{Girl}) = 80$。

$$\begin{aligned} P(\text{Gril}|\text{Pants}) &= \frac{\text{穿裤子的女生数}}{\text{穿裤子的总数}} \\ &= \frac{100 \times P(\text{Girl}) \times P(\text{Pants}|\text{Girl})}{100 \times P(\text{Boy}) \times P(\text{Pants}|\text{Boy}) + 100 \times P(\text{Girl}) \times P(\text{Pants}|\text{Girl})} \\ &= \frac{P(\text{Girl}) \times P(\text{Pants}|\text{Girl})}{P(\text{Boy}) \times P(\text{Pants}|\text{Boy}) + P(\text{Girl}) \times P(\text{Pants}|\text{Girl})} \end{aligned}$$

$$P(\text{Girl}|\text{Pants}) = \frac{P(\text{Girl}) \times P(\text{Pants}|\text{Girl})}{P(\text{Pants})}$$

$$= \frac{P(\text{Girl}) \times P(\text{Pants}|\text{Girl})}{P(\text{Pants}|\text{Boy}) \times P(\text{Boy}) + P(\text{Pants}|\text{Girl}) \times P(\text{Girl})}$$

推测穿裤子的学生是女生的概率：

$$P(\text{Girl}|\text{Pants}) = \frac{P(\text{Girl}) \times P(\text{Pants}|\text{Girl})}{P(\text{Pants})}$$

$$= \frac{P(\text{Girl}) \times P(\text{Pants}|\text{Girl})}{P(\text{Pants}|\text{Boy}) \times P(\text{Boy}) + P(\text{Pants}|\text{Girl}) \times P(\text{Girl})}$$

$$= \frac{0.4 \times 0.5}{0.4 \times 0.5 + 0.6 \times 1} = 0.25$$

下面推测这个穿裤子的学生是男生还是女生。

推测是女生的概率：

$$P(\text{Girl}|\text{Pants}) = \frac{P(\text{Girl}) \times P(\text{Pants}|\text{Girl})}{P(\text{Pants})}$$

$$= \frac{P(\text{Girl}) \times P(\text{Pants}|\text{Girl})}{P(\text{Pants}|\text{Boy}) \times P(\text{Boy}) + P(\text{Pants}|\text{Girl}) \times P(\text{Girl})}$$

$$= \frac{0.4 \times 0.5}{0.4 \times 0.5 + 0.6 \times 1} = 0.25$$

推测是男生的概率：

$$P(\text{Boy}|\text{Pants}) = \frac{P(\text{Boy}) \times P(\text{Pants}|\text{Boy})}{P(\text{Pants})}$$

$$= \frac{P(\text{Boy}) \times P(\text{Pants}|\text{Boy})}{P(\text{Pants}|\text{Boy}) \times P(\text{Boy}) + P(\text{Pants}|\text{Girl}) \times P(\text{Girl})}$$

$$= \frac{0.6 \times 1}{0.4 \times 0.5 + 0.6 \times 1} = 0.75$$

判断哪个概率大？

$$P(\text{Girl}|\text{Pants}) \overset{<}{\underset{>}{?}} P(\text{Boy}|\text{Pants}) \xrightarrow{\text{判断}} 男生$$

$$P(\text{Boy}|\text{Pants}) \times P(\text{Pants}) = P(\text{Pants}|\text{Boy}) \times P(\text{Boy})$$

6.6 最大熵原理

最大熵原理认为，在所有可能的概率模型（分布）中，熵最大的模型是最好的模型（分布）。注意：我们通常在满足约束条件的模型集合中选取熵最大的模型。

假设离散型随机变量 X 的概率分布为 $P(X)$，则其熵为

$$H(P) = -\sum_x P(x)\log P(x)$$

假设分类模型是一个条件概率分布，X 表示输入，Y 表示输出。

注意：这个模型表示的是对于给定的输入 X，以条件概率输出 Y。

给定一个训练数据集

$$T = \{(x_1, y_1), (x_2, y_2), \cdots, (x_N, y_N)\}$$

注意：目标是用最大熵原理选择最好的分类模型。

在给定的训练集中，可以确定联合分布 $P(X, Y)$ 和边缘分布 $P(X)$ 的经验分布，分别用 $\widetilde{P}(X,Y)$ 和 $\widetilde{P}(X)$ 表示。

$$\widetilde{P}(X=x, Y=y) = \frac{v(X=x, Y=y)}{N}$$

$$\widetilde{P}(X=x) = \frac{v(X=x)}{N}$$

式中，$v(X=x, Y=y)$ 表示样本 (x, y) 出现的频数；$v(X=x)$ 表示 x 出现的频数；N 表示训练样本容量。

6.6.1 最大熵模型

最大熵分类器模型是朴素贝叶斯分类器模型的泛化。像朴素贝叶斯模型一样，最大熵分类器为给定的输入值计算每个标签的可能性，通过将适合于输入值和标签的参数乘在一起。朴素贝叶斯分类器模型为每个标签定义一个参数，指定其先验概率，为每个(特征,标签)对定义一个参数，指定独立的特征对一个标签的可能性的贡献。

相比之下，最大熵分类器模型留给用户来决定什么样的标签和特征组合应该得到自己的参数。特别地，它可以使用一个单独的参数关联一个特征与一个以上的标签；或者关联一个以上的特征与一个给定的标签。这有时会允许模型"概括"相关的标签或特征之间的一些差异。

每个接收它自己的参数的标签和特征的组合被称为一个联合特征。注意：联合特征是有标签的值的属性，而（简单）特征是未加标签的值的属性。

描述和讨论最大熵模型的文字中，术语"特征 features"往往指联合特征；术语"上下文 contexts"指（简单）特征。

通常情况下，用来构建最大熵模型的联合特征完全镜像朴素贝叶斯模型使用的联合特征。特别地，每个标签定义的联合特征对应于 w[label]，每个（简单）特征和标签组合定义的联合特征对应于 w[f, label]。给定一个最大熵模型的联合特征，分配到一个给定输入的标签的得分与适用于该输入和标签的联合特征相关联的参数的简单的乘积。

```
P(input, label) = ∏joint-features(input, label)w[joint-feature]
```

6.6.2 熵的最大化

进行最大熵分类的直觉是我们应该建立一个模型,捕捉单独的联合特征的频率,不必做任何无根据的假设。

假设我们被分配从 10 个可能的任务的列表(标签从 A-J)中为一个给定的词找出正确词义的任务。首先,我们没有被告知其他任何关于词或词义的信息。我们可以为 10 种词义选择的概率分布很多,例如:

	A	B	C	D	E	F	G	H	I	J
(i)	10%	10%	10%	10%	10%	10%	10%	10%	10%	10%
(ii)	5%	15%	0%	30%	0%	8%	12%	0%	6%	24%
(iii)	0%	100%	0%	0%	0%	0%	0%	0%	0%	0%

虽然这些分布都可能是正确的,我们很可能会选择分布(i),因为没有任何更多的信息,也没有理由相信任何词的词义比其他的更有可能。另一方面,分布(ii)及(iii)反映的假设不被我们已知的信息支持。

直觉上这种分布(i)比其他的更"公平",解释这个的一个方法是引用熵的概念。在决策树的讨论中,我们描述了熵作为衡量一套标签是如何"无序"。特别地,如果是一个单独的标签则熵较低,但如果标签的分布比较均匀则熵较高。在我们的例子中,我们选择了分布(i)因为它的标签概率分布均匀——换句话说,因为它的熵较高。一般情况下,最大熵原理是说在与我们所知道的一致的分布中,我们会选择熵最高的。

接下来,假设我们被告知词义 A 出现的次数占 55%。再一次,有许多分布与这一条新信息一致,例如:

	A	B	C	D	E	F	G	H	I	J
(iv)	55%	45%	0%	0%	0%	0%	0%	0%	0%	0%
(v)	55%	5%	5%	5%	5%	5%	5%	5%	5%	5%
(vi)	55%	3%	1%	2%	9%	5%	0%	25%	0%	0%

但是,我们可能会选择最少无根据的假设的分布——在这种情况下,分布(v)。

最后,假设我们被告知词 up 出现在 nearby 上下文中的次数占 10%,当它出现在这个上下文中时有 80%的可能使用词义 A 或 C。从这个意义上讲,将使用 A 或 C。在这种情况下,我们很难手工找到合适的分布;然而,可以验证下面看起来适当的分布。

		A	B	C	D	E	F	G	H	I	J
(vii)	+up	5.1%	0.25%	2.9%	0.25%	0.25%	0.25%	0.25%	0.25%	0.25%	0.25%
	-up	49.9%	4.46%	4.46%	4.46%	4.46%	4.46%	4.46%	4.46%	4.46%	4.46%

特别地,与我们所知道的一致的分布:将 A 列的概率加起来是 55%,将第 1 行的概率加起来是 10%;将+up 行词义 A 和 C 的概率加起来是 8%(或+up 行的 80%)。此外,其余的概率"均匀分布"。

纵观这个例子,我们将自己限制在与我们所知道的一致的分布上。其中,我们选择最高熵的分布。这正是最大熵分类器所做的。特别地,对于每个联合特征,最大熵模型计算该特征的"经验频率"——即它出现在训练集中的频率。然后,它搜索熵最大的分布,同时也预测每个联合特征正确的频率。

6.6.3 生成式分类器对比条件式分类器

朴素贝叶斯分类器和最大熵分类器之间的一个重要差异是它们可以被用来回答问题的类型。朴素贝叶斯分类器是一个生成式分类器的例子,建立一个模型,预测 P(input,label),即(input,label)对的联合概率。因此,生成式模型可以用来回答下列问题。

(1)一个给定输入的最可能的标签是什么?
(2)对于一个给定输入,一个给定标签有多大可能性?
(3)最有可能的输入值是什么?
(4)一个给定输入值的可能性有多大?
(5)一个给定输入具有一个给定标签的可能性有多大?
(6)对于一个可能有两个值中的一个值(但我们不知道是哪个)的输入,最可能的标签是什么?

最大熵分类器是条件式分类器的一个例子。条件式分类器建立模型预测 P(label|input)——一个给定输入值的标签的概率。因此,条件式模型仍然可以被用来回答问题(1)和问题(2)。然而,条件式模型不能用来回答剩下的问题(3)~问题(6)。

一般情况下,生成式模型确实比条件式模型强大,因为可以从联合概率 P(input,label)计算出条件概率 P(label|input),但反过来不行。然而,这种额外的能力是要付出代价的。该模型更强大,也有更多的"自由参数"需要学习,而训练集的大小是固定的,因此,使用一个更强大的模型时,我们可用来训练每个参数的值的数据也更少,使其难以找到最佳参数值。结果是一个生成式模型回答问题(1)和问题(2)可能不会与条件式模型一样好,因为条件式模型可以集中精力在这两个问题上。然而,如果我们确实需要像问题(3)~问题(6)的答案,那么我们别无选择,只能使用生成式模型。

生成式模型与条件式模型之间的差别类似于一张地形图和一张地平线的图片之间的区别。虽然地形图可用于回答问题更广泛,制作一张精确的地形图也明显比制作一张精确的地平线图片更加困难。

小 结

(1)为语料库中的语言数据建模可以帮助我们理解语言模型,也可以用于预测新语言数据。

（2）有监督分类器使用加标签的训练语料库来建立模型，基于输入的特征，预测那个输入的标签。

（3）有监督分类器可以执行多种 NLP 任务，包括文档分类、词性标注、语句分割、对话行为类型识别以及确定蕴含关系和很多其他任务。

（4）训练一个有监督分类器时，应该把语料分为三个数据集：用于构造分类器模型的训练集、用于帮助选择和调整模型特性的开发测试集，以及用于评估最终模型性能的测试集。

（5）评估一个有监督分类器时，重要的是要使用新鲜的没有包含在训练集或开发测试集中的数据。否则，评估结果可能会不切实际。

（6）决策树可以自动地构建树结构的流程图，用于为输入变量值基于它们的特征加标签，虽然它们易于解释，但不适合处理特性值在决定合适标签过程中相互影响的情况。

（7）在朴素贝叶斯分类器中，每个特征决定应该使用哪个标签的贡献是独立的。它允许特征值间有关联，但当两个或更多的特征高度相关时将会有问题。

（8）最大熵分类器使用的基本模型与朴素贝叶斯相似；不过，它们使用了迭代优化来寻找使训练集的概率最大化的特征权值集合。

（9）大多数从语料库自动构建的模型都是描述性的，也就是说，它们让我们知道哪些特征与给定的模式或结构相关，但它们没有给出关于这些特征和模式之间的因果关系的任何信息。

练 习

1. 阅读在本章中提到的语言技术（如词义消歧、语义角色标注、问答系统、机器翻译，命名实体识别（ne）等）之一，找出开发这种系统需要什么类型和多少数量的已标注的数据，为什么你会认为需要大量的数据？

对于有少量平衡的标签和一个多样化的测试集的分类任务，只要 100 个评估实例就可以进行有意义的评估。但是，如果一个分类任务有大量的标签或包括罕见的标签，那么选择的测试集的大小就要保证出现次数最少的标签至少出现 50 次。此外，如果测试集包含许多密切相关的实例，例如，来自一个单独文档中的实例，那么测试集的大小应增加，以确保这种多样性的缺乏不会扭曲评估结果。

2. 使用任何本章所述的三种分类器之一以及你能想到的特征，尽量好地建立一个名字性别分类器。从将名字语料库分成 3 个子集开始：500 个词为测试集，500 个词为开发测试集，剩余 6 900 个词为训练集。然后从示例的名字性别分类器开始，逐步改善。使用开发测试集检查你的分类器。一旦你对你的分类器感到满意，在测试集上检查它的最终性能。相比在开发测试集上的性能，它在测试集上的性能如何？这是你期待的吗？

```
from nltk.corpus import names
from random import shuffle
from nltk import NaiveBayesClassifier
from nltk.classify import accuracy
def gender_features(word):
    return {'last_letter':word[-1]}
```

```
names=[(name, 'male') for name in names.words('male.txt')]+[(name,
'female') for name in names.words('female.txt')]
shuffle(names)
featuresets=[(gender_features(n), g) for n, g in names]
test_set, train_set=featuresets[:500], featuresets[500:]
classifier=NaiveBayesClassifier.train(train_set)
print accuracy(classifier, test_set)
```

3. Senseval 2 语料库包含旨在训练词-词义消歧分类器的数据。它包含 4 个词的数据：hard、interest、line 和 serve。选择这 4 个词中的一个，加载相应的数据。

```
from nltk.corpus import senseval
instances=senseval.instances('hard.pos')
size=int(len(instances)×0.1)
train_set, test_set=instances[size:], instances[:size]
```

使用这个数据集，建立一个分类器，预测一个给定的实例的正确的词义标签。

4. 使用本章讨论过的电影评论文档分类器，产生对分类器最有信息量的 30 个特征的列表。你能解释为什么这些特定特征具有信息量吗？你能在它们中找到什么惊人的发现吗？

5. 选择一个本章所描述的分类任务，如名字性别检测、文档分类、词性标注或对话行为分类。使用相同的训练和测试数据，相同的特征提取器，建立该任务的三个分类器：决策树、朴素贝叶斯分类器和最大熵分类器。比较你所选任务上这三个分类器的性能。你如何看待如果使用了不同的特征提取器，三者的结果可能会不同？

```
#词性标注, brown, common_suffixes, 决策树
from nltk.corpus import brown
import nltk
suffix_fdist=nltk.FreqDist()
for word in brown.words():
    word=word.lower()
    suffix_fdist.inc(word[-1:])
    suffix_fdist.inc(word[-2:])
    suffix_fdist.inc(word[-3:])

common_suffixes=suffix_fdist.keys()[:100]
def pos_features(word):
    features={}
    for suffix in common_suffixes:
        features['endswith(%s)' % suffix]=word.lower().endswith(suffix)
    return features
```

```
tagged_words=brown.tagged_words(categories='news')[:1000]
featuresets=[(pos_features(n), g) for n, g in tagged_words]
size=int(len(featuresets)×0.1)
test_set, train_set=featuresets[:size], featuresets[size:]
classifier=nltk.DecisionTreeClassifier.train(train_set)
nltk.classify.accuracy(classifier, test_set) #0.65
#词性标注, brown, common_suffixes, 朴素贝叶斯分类器
from nltk.corpus import brown
import nltk
suffix_fdist=nltk.FreqDist()
for word in brown.words():
    word=word.lower()
    suffix_fdist.inc(word[-1:])
    suffix_fdist.inc(word[-2:])
    suffix_fdist.inc(word[-3:])

common_suffixes=suffix_fdist.keys()[:100]
def pos_features(word):
    features={}
    for suffix in common_suffixes:
        features['endswith(%s)' % suffix]=word.lower().endswith(suffix)
    return features

tagged_words=brown.tagged_words(categories='news')[:1000]
featuresets=[(pos_features(n), g) for n, g in tagged_words]
size=int(len(featuresets)×0.1)
test_set, train_set=featuresets[:size], featuresets[size:]
classifier=nltk.NaiveBayesClassifier.train(train_set)
nltk.classify.accuracy(classifier, test_set) #0.55
classifier.show_most_informative_features()
```

6. 同义词 strong 和 powerful 的模式不同（尝试将它们与 chip 和 sales 结合）。哪些特征与这种区别有关？建立一个分类器，预测每个词何时该被使用。

7. 对话行为分类器为每个帖子分配标签，不考虑帖子的上下文背景。然而，对话行为是高度依赖上下文的，一些对话行序列可能比别的更相近。例如，ynQuestion 对话行为更容易被一个 yanswer 回答而不是以一个问候来回答。利用这一事实，建立一个连续的分类器，为对话行为加标签（一定要考虑哪些特征可能是有用的，参见例 6-5 的词性标记的连续分类器的代码）。

8. 词特征在处理文本分类中是非常有用的，因为在一个文档中出现的词对于其语义内容是什么具有强烈的指示作用。然而，很多词很少出现，一些在文档中的最有信息量的词可能永远不会出现在训练数据中。一种解决方法是使用一个词典，它描述了词之间的不同。使用WordNet 词典，加强本章介绍的电影评论文档分类器，使用概括一个文档中出现的词的特征，使之更容易匹配在训练数据中发现的词。

9. PP 附件语料库是描述介词短语附着决策的语料库。语料库中的每个实例被编码为 PPAttachment 对象。

```
>>>from nltk.corpus import ppattach
>>>ppattach.attachments('training')
[PPAttachment(sent='0', verb='join', noun1='board',
prep='as', noun2='director', attachment='V'),
PPAttachment(sent='1', verb='is', noun1='chairman',
prep='of', noun2='N.V.', attachment='N'),
...]
>>>inst=ppattach.attachments('training')[1]
>>>(inst.noun1, inst.prep, inst.noun2)
('chairman', 'of', 'N.V.')
```

选择 inst.attachment 为 N 的唯一实例。

```
>>>nattach=[inst for inst in ppattach.attachments('training')
...if inst.attachment=='N']
```

使用此子语料库，建立一个分类器，尝试预测哪些介词是用来连接一对给定的名词。例如给定的名词对 team 和 researchers，分类器应该预测出介词 of。

10. 假设你想自动生成一个场景的散文描述，每个实体已经有了一个唯一描述此实体的词，如 the bo，只是想决定在有关的各项目中是否使用 in 或 on，例如 the bo is in the cupboard 对 the bo is on the shelf。通过查找语料数据，编写需要的程序，探讨这个问题，并思考下面的例子。

```
a.in the car versus on the train
b.in town versus on campus
c.in the picture versus on the screen
d. in Macbeth versus on Letterman
```

第 7 章 信息提取

7.1 信息提取概述

信息有很多种形式,一个重要的形式是结构化数据。例如,给定一个公司,希望确定它做业务的位置,反过来,给定位置,确定哪些公司在该位置做业务。数据可以是表格形式,见表 7-1。

表 7-1 locs

OrgName	LocationName
Omnicom	New York
DDB Needham	New York
Kaplan Thaler Group	New York
BBDO South	Atlanta
Georgia-Pacific	Atlanta

哪些公司在亚特兰大经营呢?可以利用查询工具 SQL 进行查询。

```sql
create table locs(OrgName varchar(255) primary key, LocationName varchar(255));
insert into locs values('Omnicom', 'New York');
insert into locs values('DDB Needham', 'New York');
insert into locs values('Kaplan Thaler Group', 'New York');
insert into locs values('BBDO South', 'Atlanta');
insert into locs values('Georgia-Pacific', 'Atlanta');
select OrgName
from locs
where LocationName='Atlanta';
```

如果这个 locs 数据作为一个元组(entity, relation, entity)列表存储在 Python 中,那么,哪些公司在亚特兰大经营呢?利用列表推导:

```
>>> locs=[
... ('Omnicom', 'IN', 'New York'),
... ('DDB Needham', 'IN', 'New York'),
```

```
...   ('Kaplan Thaler Group', 'IN', 'New York'),
...   ('BBDO South', 'IN', 'Atlanta'),
...   ('Georgia-Pacific', 'IN', 'Atlanta')
... ]
>>> query=[e1 for (e1, _, e2) in locs if e2=='Atlanta']
>>> print query
['BBDO South', 'Georgia-Pacific']
```

思考下面的片段：

> The fourth Wells account moving to another agency is the packaged paper-products division of Georgia-Pacific Corp., which arrived at Wells only last fall. Like Hertz and the History Channel, it is also leaving for an Omnicom-owned agency, the BBDO South unit of BBDO Worldwide. BBDO South in Atlanta, which handles corporate advertising for Georgia-Pacific, will assume additional duties for brands like Angel Soft toilet tissue and Sparkle paper towels, said Ken Haldin, a spesman for Georgia-Pacific in Atlanta.

通读上文，可以收集到回答例子问题所需的信息，但如何能让一台机器理解上文返回列表['BBDO South', 'Georgia-Pacific']作为答案呢？这显然是一个更困难的任务，与表7.1不同，上文不包含连结公司名和位置名的结构。这个问题的解决方法之一是建立一个非常通用的意义重现。本章采取不同的方法，只查找文本中非常具体的各种信息，如公司和地点之间的关系，不是试图用文字像上文那样直接回答这个问题。

首先，将自然语言句子这样的非结构化数据转换成表7-1的结构化数据，然后，利用查询工具SQL，这种从文本获取意义的方法称为信息提取。信息提取有许多应用，包括商业智能、媒体分析、情感检测、专利检索、电子邮件扫描。当前研究的一个特别重要的领域是提取出电子科学文献的结构化数据，特别是在生物学和医学领域。

一个简单的信息提取系统，首先，要使用句子分割器将该文档的原始文本分割成句，使用分词器将每个句子进一步细分为词。接下来，对每个句子进行词性标注，在下一步命名实体识别中将证明这是非常有益的。在这一步，寻找每个句子中潜在的实体。最后，使用关系识别搜索文本中不同实体间的可能关系。该系统以一个文档的原始文本作为其输入，生成(entity, relation, entity)元组的列表作为输出，例如，假设一个文档表明Georgia-Pacific公司位于Atlanta，它可能产生元组([ORG: 'Georgia-Pacific'], 'in', [LOC: 'Atlanta'])。可以定义一个函数，简单连接NLTK中默认的句子分割器、分词器和词性标注器。

```
Python 2.7.11 |Anaconda 2.5.0 (64-bit)| (default, Jan 29 2016,
14:26:21) [MSC v.1500 64 bit (AMD64)] on win32
>>> from nltk import sent_tokenize, word_tokenize, pos_tag
>>> def ie_preprocess(document):
...     sentences=sent_tokenize(document)
```

```
...         #['Excuse me .', 'Is there any summer resort around the town ?']
...         #sentences(list of string)
...         sentences=[word_tokenize(sent) for sent in sentences]
...         #[['Excuse', 'me', '.'], ['Is', 'there', 'any', 'summer', 'resort', 'around', 'the', 'town', '?']]
...         #tokenized sentences(list of lists of strings)
...         sentences=[pos_tag(sent) for sent in sentences]
...         #[[('Excuse', 'IN'), ('me', 'PRP'), ('.', '.')], [('Is', 'VBZ'), ('there', 'EX'), ('any', 'DT'), ('summer', 'NN'), ('resort', 'NN'), ('around', 'IN'), ('the', 'DT'), ('town', 'NN'), ('?', '.')]]
...         #part of speech tagging
...         print sentences
...
>>> document='Excuse me . Is there any summer resort around the town ?'
>>> #raw text(string)
... ie_preprocess(document)
[[('Excuse', 'IN'), ('me', 'PRP'), ('.', '.')], [('Is', 'VBZ'), ('there', 'EX'), ('any', 'DT'), ('summer', 'NN'), ('resort', 'NN'), ('around', 'IN'), ('the', 'DT'), ('town', 'NN'), ('?', '.')]]
>>> #raw text(string)
... ie_preprocess(document)
Traceback (most recent call last):
  File "<stdin>", line 2, in <module>
  File "<stdin>", line 8, in ie_preprocess
  File "C:\Anaconda2\lib\site-packages\nltk\tag\__init__.py", line 110, in pos_tag
    tagger = PerceptronTagger()
  File "C:\Anaconda2\lib\site-packages\nltk\tag\perceptron.py", line 142, in __init__
    self.load(AP_MODEL_LOC)
  File "C:\Anaconda2\lib\site-packages\nltk\tag\perceptron.py", line 210, in load
    with open(loc, 'rb') as fin:
IOError: [Errno 2] No such file or directory: 'C:\\nltk_data\\taggers\\averaged_perceptron_tagger.zip\\averaged_perceptron_tagger\\averaged_perceptron_tagger.pickle'
```

有些 nltk3.1 不会自动解压，解决方法是手工解压 C:\nltk_data\taggers\averaged_perceptron_tagger.zip 即可。

```
>>> from nltk.tag import pos_tag
>>> from nltk.tokenize import word_tokenize
>>> pos_tag(word_tokenize("John's big idea isn't all that bad."))
[('John', 'NNP'), ("'s", 'POS'), ('big', 'JJ'), ('idea', 'NN'),
('is', 'VBZ'), ("n't", 'RB'), ('all', 'PDT'), ('that', 'DT'), ('bad',
'JJ'), ('.', '.')]
>>> pos_tag(word_tokenize("John's big idea isn't all that bad."),
tagset='universal')
[('John', u'NOUN'), ("'s", u'PRT'), ('big', u'ADJ'), ('idea',
u'NOUN'), ('is', u'VERB'), ("n't", u'ADV'), ('all', u'DET'), ('that',
u'DET'), ('bad', u'ADJ'), ('.', u'.')]
```

接下来，命名实体识别中，分割和标注可能组成一个有趣关系的实体。通常情况下，这些将被定义为名词短语。在一些任务中，同时考虑不明确的名词或名词块也是有用的，这些不必要一定与定义 NP 和适当名称一样的方式指示实体，最后在提取关系时，搜索对文本中出现在附近的实体对之间的特殊模式，并使用这些模式建立元组记录实体之间的关系。

7.2 分 块

我们将用于实体识别的基本技术称为分块（chunking），文法抽象过程：

```
'We saw the yellow dog'
['We', 'saw', 'the', 'yellow', 'dog']
[PRP, VBD, DT, JJ, NN]
[[PRP], VBD, [DT, JJ, NN]]
[NP, VBD, NP]
```

本节将在较深的层面探讨分块，以块的定义和表示开始，将看到正则表达式和 N-gram 的方法分块，使用 CoNLL-2000 分块语料库开发和评估分块器。

7.2.1 名词短语分块

先思考名词短语分块或 NP 分块（NP chunking）。在这里我们寻找（识别）单独名词短语对应的块，其中 NP 块用方括号标记。

```
[The/DT market/NN] for/IN [system management/NN software/NN] for/IN
[Digital/NNP] ['s/POS hardware/NN] is/VBZ fragmented/JJ enough/RB
that/IN [a/DT giant/NN] such/JJ as/IN [Computer/NNP Associates/NNPS]
should/MD do/VB well/RB there/RB ./.
```

NP 块往往是比完整名词短语小的片段，例如，the market for system management software for Digital's hardware 是一个单独的名词短语，含两个嵌套的名词短语，它中间有一个简单的 NP 块 the market，这种差异是 NP 块被定义为不包含其他的 NP 块（素短语）。因此，修饰一个名词的任何介词短语或从句将不包括在相应的 NP 块内，因为它们几乎可以肯定包含更多的名词短语。

NP 分块信息最有用的来源之一是词性标记，这是在信息提取系统中进行词性标注的目的之一。在例 7-1 中用一个已经标注词性的例句来演示这种方法，为了创建一个 NP 块，先定义一个块语法，在本例中，用一个正则表达式定义一个简单的语法。这条规则是，一个 NP 块由一个可选的限定词（DT）后面跟任意数目的形容词（JJ）然后是一个名词（NN）组成（语言定义），使用此语法，我们创建了一个块分析器，分析例句，建立一棵树。

【例 7-1】 一个简单的基于正则表达式的 NP 分块器。

```
from nltk import RegexpParser
sentence=[("the","DT"),("little","JJ"),("yellow","JJ"),("dog","NN"),("barked","VBD"),("at","IN"),("the","DT"),("cat","NN")]
#二元组单词序列
grammar="NP: {<DT>?<JJ>*<NN>}" #名词短语的文法
cp=RegexpParser(grammar) #语法分析器
result=cp.parse(sentence) #语法分析
print result
result.draw() #语法分析树
```

广义表示的语法树：

```
>>> from nltk import RegexpParser
>>> sentence=[("the", "DT"), ("little", "JJ"), ("yellow", "JJ"), ("dog", "NN"), ("barked", "VBD"), ("at", "IN"), ("the", "DT"), ("cat", "NN")] #二元组单词序列
>>> grammar="NP: {<DT>?<JJ>*<NN>}" #名词短语的文法
>>> cp=RegexpParser(grammar) #语法分析器
>>> result=cp.parse(sentence) #语法分析
>>> print result
(S
  (NP the/DT little/JJ yellow/JJ dog/NN)
  barked/VBD
  at/IN
  (NP the/DT cat/NN))
>>> result.draw() #语法分析树
```

语法树：

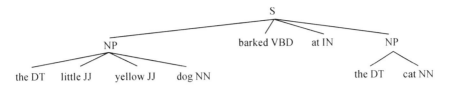

7.2.2 标记模式

块的语法规则使用标记模式来描述,一个标记模式是一个用尖括号分隔的词性标记序列,如{<DT>?<JJ.*>*<NN.*>+},匹配以一个可选的限定词开头,后面跟零个或多个任何类型的形容词后面跟一个或多个任何类型名词的标识符序列。

```
another/DT sharp/JJ dive/NN
trade/NN figures/NNS
any/DT new/JJ policy/NN measures/NNS
earlier/JJR stages/NNS
Panamanian/JJ dictator/NN Manuel/NNP Noriega/NNP
```

使用图形界面 from nltk.app import chunkparser;chunkparser()测试它们,如图 7-1 所示。

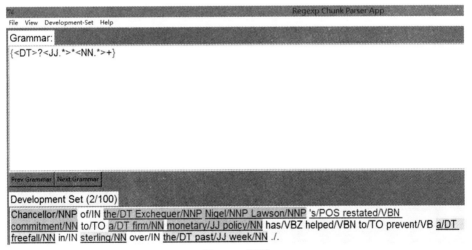

图 7-1 测试

7.2.3 用正则式分块

要找到一个给定的句子的块结构,RegexpParser 分块器以一个没有标识符分块的平面结构开始,轮流应用分块规则,依次更新块结构,所有的规则都被调用后,返回块结构。

例 7-2 显示了一个由两条规则组成的块语法。第一条规则匹配一个可选的限定词或所有格代名词,零个或多个形容词,然后跟一个名词。第二条规则匹配一个或多个专有名词。还定义了一个进行分块的例句,并在此输入上运行这个分块器。

【例7-2】 简单的名词短语分块器。

```
>>> from nltk import RegexpParser
>>> grammar="""
...     NP:{<DT|PP\$>?<JJ>*<NN>}
...     #chunk determiner/possessive, adjectives and nouns
...     {<NNP>+}
...     #chunk sequences of proper nouns
... """
>>> cp=RegexpParser(grammar)
>>> sentence=[("Rapunzel", "NNP"), ("let", "VBD"), ("down", "RP"),
("her", "PP$"), ("long", "JJ"), ("golden", "JJ"), ("hair", "NN")]
>>> print cp.parse(sentence)
(S
  (NP Rapunzel/NNP)
  let/VBD
  down/RP
  (NP her/PP$ long/JJ golden/JJ hair/NN))
>>> cp.parse(sentence).draw()
```

```
>>> cp.parse(sentence).draw()
>>> from nltk import RegexpParser
>>> grammar="""
...     A:{<ID><\$ASSIGN><E>*}
... """
>>> cp=RegexpParser(grammar)
>>> sentence=[("x1", "ID"), ("=", "$ASSIGN"), ("2", "E")]
>>> print cp.parse(sentence)
(S (A x1/ID =/$ASSIGN 2/E))
>>> cp.parse(sentence).draw()
```

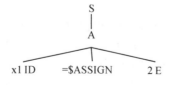

$符号是正则式中的一个特殊字符,必须使用转义符\来匹配 PP$标记。如果标记模式匹

配位置重叠，最左边的匹配优先，例如，如果我们应用一个匹配两个连续的名词文本的规则到一个包含三个连续名词的文本，则只有前两个名词将分块。

```
nouns=[("money", "NN"), ("market", "NN"), ("fund", "NN")]
grammar="NP:{<NN><NN>} #Chunk two consecutive nouns"
cp=RegexpParser(grammar)
print cp.parse(nouns)
cp.parse(nouns).draw()
```

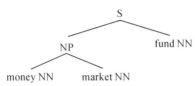

一旦创建了块 money market，就已经消除了允许 fund 被包含在一个块中的上下文。这个问题可以避免，使用一种更加宽容的块规则，如 NP:{<NN>+}。

```
nouns=[("money", "NN"), ("market", "NN"), ("fund", "NN")]
grammar="NP:{<NN>+} #Chunk two consecutive nouns"
cp=RegexpParser(grammar)
print cp.parse(nouns)
cp.parse(nouns).draw()
```

我们已经为每个块规则添加了一个注释（这些都是可选的）。

7.2.4 文本语料库探索

使用分块器寻找三词短语：

```
Python 2.7.11 |Anaconda 2.5.0 (64-bit)| (default, Jan 29 2016,
14:26:21) [MSC v.1500 64 bit (AMD64)] on win32
>>> from nltk import RegexpParser
>>> from nltk.corpus import brown
>>> flag=False
>>> cp=RegexpParser('CHUNK:{<V.*><TO><V.*>}')
>>> for sent in brown.tagged_sents():
...     tree=cp.parse(sent)
...     for subtree in tree.subtrees():
```

```
...             if subtree.label()=='CHUNK':
...                 flag=True
...                 print subtree
...                 subtree.draw()
...                 break
...         if flag:
...             break
...
(CHUNK combined/VBN to/TO achieve/VB)
```

注意 subtree.label()和 subtree.node 的变化,将上面的例子封装在函数 find_chunks()内,以一个如 CHUNK:{<V.*><TO><V.*>}的块字符串作为参数,用它来搜索语料库寻找其他几个模式,如 4 个或更多的连续的名词,即"NOUNS:{<N.*>{4, }}"。

7.2.5 加缝隙

有时定义从一个块中排除比较容易,此时,可以为不包括在一大块中的一个标识符序列定义一个缝隙。在下面的例子中,barked/VBD at/IN 是一个缝隙。

```
[the/DT little/JJ yellow/JJ dog/NN] barked/VBD at/IN [the/DT cat/NN]
```

加缝隙是从一大块中去除一个标识符序列的过程,如果匹配的标识符序列贯穿一整块,那么这一整块会被去除;如果标识符序列出现在块中间,这些标识符会被去除,在以前只有一个块的地方留下两个块;如果序列在块的周边,这些标记被去除,留下一个较小的块。表 7-2 演示了这三种可能性,例 7-3 将整个句子作为一个块。

表 7-2 三个加缝隙规则应用于同一个块

	整个块	块中间	块结尾
输入	[a/DT little/JJ dog/NN]	[a/DT little/JJ dog/NN]	[a/DT little/JJ dog/NN]
操作	Chink DT JJ NN	Chink JJ	Chink NN
模式	}DT JJ NN{	}JJ{	}NN{
输出	a/DT little/JJ dog/NN	[a/DT] little/JJ [dog/NN]	[a/DT little/JJ] dog/NN

【例 7-3】 简单的加缝器。

```
>>> from nltk import RegexpParser
>>> grammar="""NP:{<.*>+} #Chunk everything
...  }<VBD|IN>+{ #Chink sequences of VBD and IN
...  """
>>> sentence=[("the", "DT"), ("little", "JJ"), ("yellow", "JJ"),
("dog", "NN"), ("barked", "VBD"), ("at", "IN") , ("the", "DT"), ("cat",
"NN")]
```

```
>>> cp=RegexpParser(grammar)
>>> print cp.parse(sentence)
(S
  (NP the/DT little/JJ yellow/JJ dog/NN)
  barked/VBD
  at/IN
  (NP the/DT cat/NN))
>>> cp.parse(sentence).draw()
```

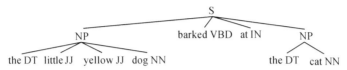

等价于:

```
>>> from nltk import RegexpParser
>>> grammar="""NP:{<DT><JJ>*<NN>} #Chunk NP
... """
>>> sentence=[("the", "DT"), ("little", "JJ"), ("yellow", "JJ"),
("dog", "NN"), ("barked
", "VBD"), ("at", "IN") , ("the", "DT"), ("cat", "NN")]
>>> cp=RegexpParser(grammar)
>>> print cp.parse(sentence)
(S
  (NP the/DT little/JJ yellow/JJ dog/NN)
  barked/VBD
  at/IN
  (NP the/DT cat/NN))
>>> cp.parse(sentence).draw()
```

7.2.6 块的表示

作为标注和分析之间的中间状态，块结构可以使用标记或树来表示，使用最广泛的表示是 IOB 标记。在这个方案中，每个标识符用三个特殊的块标签标注，I, O 或 B。块的开始标注为 B，块内的标识符标注为 I，所有其他的标识符标注为 O。B 和 I 标记可以加块类型的后缀，如 B-NP，I-NP，当然，如果没有必要指定出现在块外标识符的类型，可以只标注为 O。IOB 标记已成为文件中表示块结构的标准方式，这里将使用这种格式。在此表示中，每行由标识符、词性标记和块标记组成，这种格式允许我们表示多个块类型。注意，行首不能有空格，否则:

```
Traceback (most recent call last):
  File "<stdin>", line 1, in <module>
  File "C:\Anaconda2\lib\site-packages\nltk\chunk\util.py", line 387, in conllstr2tree
    raise ValueError('Error on line %d' % lineno)
ValueError: Error on line 1
```

可以研究 C:\Anaconda2\lib\site-packages\nltk\chunk\util.py 中的 conllstr2tree。

```
We PRP B-NP
saw VBD O
the DT B-NP
yellow JJ I-NP
dog NN I-NP
```

用文法表示：

```
S→NP VBD NP
NP→PRP|DT JJ NN
PRP→'We'
VBD→'saw'
DT→'the'
JJ→'yellow'
NN→'dog'
```

块的结构也可以使用树表示，这有利于块作为一个组成部分直接操作，NLTK 中树作为块的内部表示，并提供这些树与 IOB 格式互换的方法。

```
Python 2.7.11 |Anaconda 2.5.0 (64-bit)| (default, Jan 29 2016, 14:26:21) [MSC v.1500 64 bit (AMD64)] on win32
>>>from nltk.chunk import conllstr2tree
>>>text='''
...We PRP B-NP #行首不能有空格，否则
...saw VBD O
...the DT B-NP
...yellow JJ I-NP
...dog NN I-NP
...'''
>>>conllstr2tree(text, chunk_types=['NP'])
Tree('S', [Tree('NP', [(u'We', u'PRP')]), (u'saw', u'VBD'), Tree('NP', [(u'the', u'DT'), (u'yellow', u'JJ'), (u'dog', u'NN')])])
>>>conllstr2tree(text, chunk_types=['NP']).draw()
```

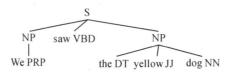

7.3 开发和评估分块

一开始是寻找将 IOB 格式转换成 NLTK 树的机制，然后是使用已分块的语料库在更大的规模上进行操作，将看到如何为一个分块器相对一个语料库的准确性打分，再看看一些数据驱动方式搜索 NP 块，重点在于扩展一个分块器的覆盖范围。

7.3.1 读取 IOB 格式与 CoNLL2000 分块语料库

使用 IOB 符号分块，每个句子使用多行表示。

```
he PRP B-NP
accepted VBD B-VP
the DT B-NP
position NN I-NP
```

转换函数 conllstr2tree()用这些多行字符串（树的定义）建立一个树表示，此外，它允许我们选择使用三个块类型的任何子集，这里只是 VP 块。

```
from nltk.chunk import conllstr2tree
text = '''
he PRP B-NP
accepted VBD B-VP
the DT B-NP
position NN I-NP
'''
conllstr2tree(text, chunk_types = ['VP']).draw()
```

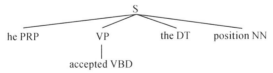

可以使用 NLTK 的 corpus 模块访问较大量的已分块文本。CoNLL2000 分块语料库分为训练和测试两部分，标注有词性标记和 IOB 格式分块标记，可以使用 nltk.corpus.conll2000 访问这些数据。下面是一个读取语料库训练部分第 100 个句子的例子。

CoNLL2000 分块语料库包含三种块类型：NP 块、VP 块（如 has already delivered）、PP 块（如 because of）。因为现在我们需要的是 NP 块，所以可以使用 chunk_types 参数选择它们。

```
from nltk.corpus import conll2000
print conll2000.chunked_sents('train.txt', chunk_types=['NP'])[99].draw()
```

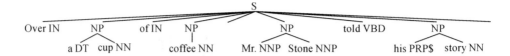

7.3.2 简单评估和基准

访问一个已分块语料库,评估分块器,为不创建任何块的块分析器 cp 建立一个基准。

```
>>> from nltk import RegexpParser
>>> from nltk.corpus import conll2000
>>> cp=RegexpParser("")
>>> test_sents=[conll2000.chunked_sents('test.txt', chunk_types=['NP'])[0]]
>>> print cp.evaluate(test_sents)
ChunkParse score:
    IOB Accuracy:  28.6%
    Precision:      0.0%
    Recall:         0.0%
    F-Measure:      0.0%
```

IOB 标记准确性表明超过 1/3 的词被标注为 O,即没有在 NP 块中,然而,由于标注器没有找到任何块,其精度、召回率(Recall)和 F-度量均为零。现在让我们尝试一个初级的正则式分块器,查找以名词短语标记的特征字母开头的标记,如 CD、DT 和 JJ。

```
>>> from nltk import RegexpParser
>>> from nltk.corpus import conll2000
>>> x=conll2000.chunked_sents('test.txt', chunk_types=['NP'])
>>> y=conll2000.tagged_sents('test.txt')
>>> grammar=r"NP:{<[CDJNP].*>+}"
>>> cp=RegexpParser(grammar)
>>> z=cp.parse(y[0])
>>> print cp.evaluate([x[0]])
ChunkParse score:
    IOB Accuracy:  92.9%
    Precision:     71.4%
    Recall:        55.6%
    F-Measure:     62.5%
>>> y[0]
[(u'Rockwell', u'NNP'), (u'International', u'NNP'), (u'Corp.', u'NNP'), (u"'s", u'POS'), (u'Tulsa', u'NNP'), (u'unit', u'NN'), (u'said', u'VBD'), (u'it', u'PRP'), (u'signed', u'VBD'), (u'a', u'DT'),
```

```
(u'tentative', u'JJ'), (u'agreement', u'NN'), (u'extending', u'VBG'),
(u'its', u'PRP$'), (u'contract', u'NN'), (u'with', u'IN'), (u'Boeing',
u'NNP'), (u'Co.', u'NNP'), (u'to', u'TO'), (u'provide', u'VB'),
(u'structural', u'JJ'), (u'parts', u'NNS'), (u'for', u'IN'),
(u'Boeing', u'NNP'), (u"'s", u'POS'), (u'747', u'CD'), (u'jetliners',
u'NNS'), (u'.', u'.')]
```

这种方法能取得相当好的结果。我们可以采用更多数据驱动的方法改善它，在这里我们使用训练语料找到对每个词性标记最有可能的块标记 I、O 或 B。换句话说，可以使用 unigram 标注器建立一个分块器，但不是尝试确定每个词正确的词性标记，而是给定每个词的词性标记，尝试确定正确的块标记。

在例 7-4 中，定义了 UnigramChunker 类，使用 unigram 标注器给句子加块标记。这个类的大部分代码只是用来在 NLTK 的 ChunkParserI 接口使用的分块树表示和嵌入式标注器使用的 IOB 表示之间镜像转换。类定义了两个方法：一个构造函数，在建立一个新的 UnigramChunker 时调用；一个 parse 方法，用来给新句子分块。

【例 7-4】 使用 unigram 标注器对名词短语分块。

```
class UnigramChunker(nltk.ChunkParserI):
    def __init__(self, train_sents):
        train_data=[[(t, c) for w, t, c in nltk.chunk.tree2conlltags(sent)]
for sent in train_sents]
        self.tagger=nltk.UnigramTagger(train_data)
    def parse(self, sentence):
        pos_tags=[pos for (word, pos) in sentence]
        tagged_pos_tags=self.tagger.tag(pos_tags)
        chunktags=[chunktag for (pos, chunktag) in tagged_pos_tags]
        conlltags=[(word, pos, chunktag) for ((word, pos), chunktag)
in zip(sentence, chunktags)]
        conlltags='\n'.join([' '.join(y) for y in conlltags])
        return nltk.chunk.conllstr2tree(conlltags)
#nltk.chunk.util.conlltags2tree#更深的地方
```

构造函数需要训练句子的一个列表，这将是块树的形式。它首先将训练数据转换成适合训练标注器的形式，使用 tree2conlltags 映射每个块树到一个（词，词性标记，块标记）三元组的列表。

使用转换好的训练数据训练一个 unigram 标注器，并存储在 self.tagger 中，供以后使用。

parse 方法取一个已标注的句子作为其输入，以从那句话提取词性标记开始。然后使用在构造函数中训练过的标注器 self.tagger，为词性标记标注 IOB 块标记。接下来，提取块标记，与原句组合，产生 conlltags。最后，使用 conllstr2tree 将结果转换成一个块树。

现在有了 UnigramChunker，可以使用 CoNLL2000 分块语料库训练它，并测试其性能。

```
import nltk
from nltk.corpus import conll2000
test_sents=conll2000.chunked_sents('test.txt',
chunk_types=['NP'])
train_sents=conll2000.chunked_sents('train.txt',
chunk_types=['NP'])
unigram_chunker=UnigramChunker(train_sents)
print unigram_chunker.evaluate(test_sents)
```

这个分块器相当不错,达到整体 F、度量 83%的得分。让我们来看一看通过使用 unigram 标注器分配一个标记给每个语料库中出现的词性标记,它学到了什么。

```
postags=sorted(set(pos for sent in train_sents
for (word, pos) in sent.leaves()))
print unigram_chunker.tagger.tag(postags)
```

它已经发现大多数标点符号出现在 NP 块外,除了两种货币符号#和$。它也发现限定词(DT)和所有格(PRP$和 WP$)出现在 NP 块的开头,而名词类型(NN, NNP, NNPS, NNS)大多出现在 NP 的块内。

建立了一个 unigram 分块器,很容易建立一个 bigram 分块器:只需要改变类的名称为 BigramChunker,修改例 7-4 中构造一个 BigramTagger 而不是 UnigramTagger。由此产生的分块器的性能略高于 unigram 分块器。

```
bigram_chunker=BigramChunker(train_sents)
print bigram_chunker.evaluate(test_sents)
```

7.3.3 训练基于分类器的分块器

无论是基于正则式的分块器还是 n-gram 分块器,决定创建什么块完全基于词性标记,然而,有时词性标记不足以确定一个句子应如何分块,例如:

```
Joey/NN sold/VBD the/DT farmer/NN rice/NN ./.
Nick/NN bre/VBD the/DT computer/NN monitor/NN ./.
```

这两句话的词性标记相同,都是"NN VBD DT NN NN .",但分块方式不同。在第一句中,the farmer 和 rice 都是单独的块,双宾语,农民种植大米,而在第二个句子中相应的部分,the computer monitor 是一个单独的块,宾语,计算机包括显示器。显然,如果想最大限度地提升分块的性能,需要使用词的内容信息作为词性标记的补充。包含词的内容信息的方法之一是使用基于分类器的分块器对句子分块,如使用的 n-gram 分块器,这个基于分类器的分块器分配 IOB 标记给句子中的词,然后将这些标记转换为块。基于分类器的 NP 分块器的基础代码如例 7-5 所示,它包括两个类,第一个类使用 MaxentClassifier,第二个类基本上是标注器类的一个包装器,将它变成一个分块器。训练期间,第二个类映射训练语料中的块树到标记序列,在 parse()方法中,它将标注器提供的标记序列转换回一个块树。

【例 7-5】 使用连续分类器对 NP 分块。

```
    Python 2.7.11 |Anaconda 2.5.0 (64-bit)| (default, Jan 29 2016,
14:26:21) [MSC v.1500 64 bit (AMD64)] on win32
    >>> from nltk import TaggerI, MaxentClassifier, ChunkParserI
    >>> from nltk.tag import untag
    >>> from nltk.chunk import tree2conlltags, conllstr2tree
    >>> class ConsecutiveNPChunkTagger(TaggerI):
    ...     def __init__(self, train_sents):
    ...         train_set=[]
    ...         for tagged_sent in train_sents:
    ...             untagged_sent=untag(tagged_sent)
    ...             history=[]
    ...             for i, (word, tag) in enumerate(tagged_sent):
    ...                 featureset=npchunk_features(untagged_sent, i, history)
    ...                 train_set.append((featureset, tag))
    ...                 history.append(tag)
    ...             self.classifier=MaxentClassifier.train(train_set, algorithm='IIS', trace=0)
    ...     def tag(self, sentence):
    ...         history=[]
    ...         for i, word in enumerate(sentence):
    ...             featureset=npchunk_features(sentence, i, history)
    ...             tag=self.classifier.classify(featureset)
    ...             history.append(tag)
    ...         return zip(sentence, history)
    ...
    >>> class ConsecutiveNPChunker(ChunkParserI):
    ...     def __init__(self, train_sents):
    ...         tagged_sents=[[((w, t), c) for w, t, c in tree2conlltags(sent)] for sent in train_sents]
    ...         self.tagger=ConsecutiveNPChunkTagger(tagged_sents)
    ...     def parse(self, sentence):
    ...         tagged_sents=self.tagger.tag(sentence)
    ...         conlltags=[(w, t, c) for ((w, t), c) in tagged_sents]
    ...         conlltags='\n'.join([' '.join(y) for y in conlltags])
    ...         return conllstr2tree(conlltags)
    #nltk.chunk.util.conlltags2tree#更深的地方
```

唯一需要填写的是特征提取器。定义一个简单的特征提取器，它只是提供了当前标识符的词性标记。使用此特征提取器，基于分类器的分块器的性能与 unigram 分块器非常类似。

```
>>> def npchunk_features(sentence, i, history):
...     word, pos=sentence[i]
...     return {"pos":pos}
...
>>> from nltk.corpus import conll2000
>>> test_sents=conll2000.chunked_sents('test.txt', chunk_types=['NP'])
>>> train_sents=conll2000.chunked_sents('train.txt', chunk_types=['NP'])
>>> chunker=ConsecutiveNPChunker(train_sents)
>>> print chunker.evaluate(test_sents)
ChunkParse score:
    IOB Accuracy:      92.9%
    Precision:         79.9%
    Recall:            86.8%
    F-Measure:         83.2%
```

还可以添加一个特征：前面词的词性标记。添加此特征允许分类器模拟相邻标记之间的相互作用，由此产生的分块器与 bigram 分块器非常接近。

```
def npchunk_features(sentence, i, history):
    word, pos=sentence[i]
    if i==0:
        prevword, prevpos="<START>", "<START>"
    else:
        prevword, prevpos=sentence[i-1]
    return {"pos":pos, "prevpos":prevpos}
```

```
chunker=ConsecutiveNPChunker(train_sents)
print chunker.evaluate(test_sents)
```

下面将尝试把当前词增加为特征，因为我们假设这个词的内容应该对分块有用。我们发现这个特征确实提高了分块器的性能约 1.5%（相应的错误率减少约 10%）。

```
def npchunk_features(sentence, i, history):
    word, pos=sentence[i]
    if i==0:
        prevword, prevpos="<START>", "<START>"
    else:
        prevword, prevpos=sentence[i-1]
    return {"pos":pos, "word":word, "prevpos":prevpos}
```

```
chunker=ConsecutiveNPChunker(train_sents)
print chunker.evaluate(test_sents)
```

尝试用多种附加特征扩展特征提取器，如预取特征、配对功能和复杂的语境特征。最后一个特征，被称为 tags-since-dt。创建一个字符串，描述自最近的限定词以来遇到的所有词性标记。

```
def npchunk_features(sentence, i, history):
    word, pos=sentence[i]
    if i==0:
        prevword, prevpos="<START>", "<START>"
    else:
        prevword, prevpos=sentence[i-1]
    if i==len(sentence)-1:
        nextword, nextpos="<END>", "<END>"
    else:
        nextword, nextpos=sentence[i+1]
    return {"pos":pos, "word":word, "prevpos":prevpos, "nextpos":nextpos,
"prevpos+pos":"%s+%s"%(prevpos, pos), "pos+nextpos":"%s+%s"%(pos, nextpos),
"tags-since-dt":tags_since_dt(sentence, i)}

def tags_since_dt(sentence, i):
    tags=set()
    for word, pos in sentence[:i]:
        if pos=='DT':
            tags=set()
        else:
            tags.add(pos)
    return'+'.join(sorted(tags))
chunker=ConsecutiveNPChunker(train_sents)
print chunker.evaluate(test_sents)
```

尝试为特征提取器函数 npchunk_features 增加不同的特征，看看是否可以进一步改善的 NP 分块器的性能。

7.4 语言结构中的递归

7.4.1 用级联分块器构建嵌套结构

创建一个包含递归规则的多级块语法，就可以建立任意深度的块结构。例 7-6 是名词短语、介词短语、动词短语和句子的模式，这是一个四级块语法器，可以用来创建深度最多为4的结构。

【例7-6】 一个分块器，可以处理NP、PP、VP和S。

```
Python 3.7.3 (default, Mar 27 2019, 17:13:21) [MSC v.1915 64 bit (AMD64)] :: Anaconda, Inc. on win32
>>> from nltk import RegexpParser
>>> grammar=r"""
... NP:{<DT|JJ|NN.*>+} #Chunk sequences of DT, JJ, NN
... PP:{<IN><NP>} #Chunk prepositions followed by NP
... VP:{<VB.*><NP|PP|CLAUSE>+$} #Chunk verbs and their arguments
... CLAUSE:{<NP><VP>} #Chunk NP, VP
... """
>>> cp=RegexpParser(grammar) #语法分析器
>>> sentence=[("Mary", "NN"), ("saw", "VBD"), ("the", "DT"), ("cat", "NN"), ("sit", "VB"), ("on", "IN"), ("the", "DT"), ("mat", "NN")]
>>> print(cp.parse(sentence)) #语法分析
(S
  (NP Mary/NN)
  saw/VBD
  (CLAUSE
    (NP the/DT cat/NN)
    (VP sit/VB (PP on/IN (NP the/DT mat/NN)))))
>>> cp.parse(sentence).draw()
```

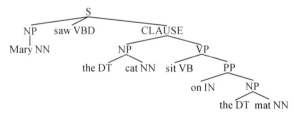

不幸的是，这一结果丢掉了以saw为首的VP（它还有其他缺陷）。将此分块器应用到一个有更深嵌套的句子时，让我们看看会发生什么。注意，它无法识别saw开始的VP块。

```
sentence=[("John", "NNP"), ("thinks", "VBZ"), ("Mary", "NN"), ("saw", "VBD"),
("the", "DT"), ("cat", "NN"), ("sit", "VB"), ("on", "IN"), ("the", "DT"), ("mat", "NN")]
print cp.parse(sentence)
cp.parse(sentence).draw()
```

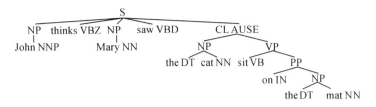

这些问题的解决方案是：让分块器在其模式中循环，尝试完所有模式之后，重复此过程。添加一个可选的第二个参数 loop 指定这套模式应该循环的次数。

```
cp=nltk.RegexpParser(grammar, loop=2)
print cp.parse(sentence)
cp.parse(sentence).draw()
```

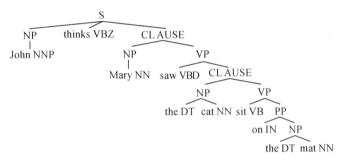

这个级联过程能创建深层结构。然而，创建和调试级联过程是困难的，关键点是它能更有效地做全面分析。另外，级联过程只能产生固定深度的树（不超过级联级数），完整的句法分析是不够的。

7.4.2 树

树是一组连接的加标签节点，从一个特殊的根节点沿一条唯一的路径到达每个节点。下面是一棵树的例子。

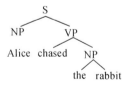

我们用家庭来比喻树中节点的关系，例如，S 是 VP 的双亲，反之，VP 是 S 的一个孩子。此外，由于 NP 和 VP 同为 S 的两个孩子，它们是兄弟。为方便起见，也有特定树的文本格式。

树可以用来编码任何同构的超越语言形式序列的层次结构。一般情况下，叶子和节点值不一定是字符串。在 NLTK 中，创建了一棵树，通过给一个节点添加标签和一个孩子列表。

```
>>> from nltk import Tree
>>> tree1=Tree('NP', ['Alice'])
>>> print tree1 #(NP Alice)
```

```
(NP Alice)
>>> tree1.draw()
```

```
tree2=nltk.Tree('NP', ['the', 'rabbit'])
tree2.draw() # (NP the rabbit)
```

可以不断合并成更大的树。

```
tree3=nltk.Tree('VP', ['chased', tree2])
```

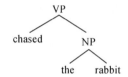

```
tree4=nltk.Tree('S', [tree1, tree3])
tree4.draw() #(S (NP Alice) (VP chased (NP the rabbit)))
```

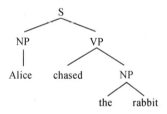

```
import nltk
tree1=nltk.Tree('flippers', ['no', 'yes'])
tree2=nltk.Tree('no surfacing', ['no', tree1])
print tree2 #(NP Alice)
tree2.draw()
```

下面是树对象的一些方法。

```
print tree4[1] #(VP chased (NP the rabbit))
```

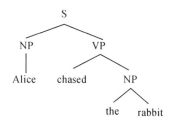

```
tree4[1].node #'VP'
tree4.leaves() #['Alice', 'chased', 'the', 'rabbit']
tree4[1][1][1] #'rabbit'
```

复杂的树用括号表示难以阅读。在这些情况下，draw 方法是非常有用的。它会打开一个新窗口，包含树的一个图形表示。树显示窗口可以放大和缩小，子树可以折叠和展开，并将图形表示输出为一个 postscript 文件（包含在一个文档中）。

```
tree3.draw()
```

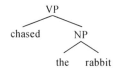

```
{'S': {0: {'NP':{0:'Alice'}}, 1: {'VP': {0: {'V':{0:'chased'}}, 1:
{'NP':{0:'the', 1:'rabbit'}}}}}}
```

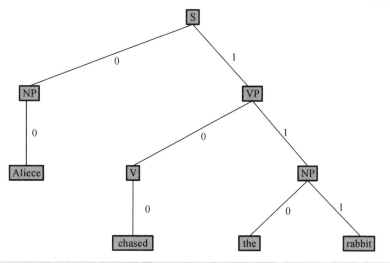

```
import matplotlib.pyplot as plt
decisionNode = dict(boxstyle="sawtooth", fc="0.8")
leafNode = dict(boxstyle="round4", fc="0.8")
arrow_args = dict(arrowstyle="<-")
def getNumLeafs(myTree):
    numLeafs = 0
    firstStr = myTree.keys()[0]
```

```python
        secondDict = myTree[firstStr]
        for key in secondDict.keys():
            if type(secondDict[key]).__name__=='dict':#test to see if the nodes are dictonaires, if not they are leaf nodes
                numLeafs += getNumLeafs(secondDict[key])
            else:   numLeafs +=1
        return numLeafs
    def getTreeDepth(myTree):
        maxDepth = 0
        firstStr = myTree.keys()[0]
        secondDict = myTree[firstStr]
        for key in secondDict.keys():
            if type(secondDict[key]).__name__=='dict':#test to see if the nodes are dictonaires, if not they are leaf nodes
                thisDepth = 1 + getTreeDepth(secondDict[key])
            else:   thisDepth = 1
            if thisDepth > maxDepth: maxDepth = thisDepth
        return maxDepth
    def plotNode(nodeTxt, centerPt, parentPt, nodeType):
        createPlot.ax1.annotate(nodeTxt, xy=parentPt, xycoords='axes fraction',
                xytext=centerPt, textcoords='axes fraction',
                va="center", ha="center", bbox=nodeType, arrowprops=arrow_args )

    def plotMidText(cntrPt, parentPt, txtString):
        xMid = (parentPt[0]-cntrPt[0])/2.0 + cntrPt[0]
        yMid = (parentPt[1]-cntrPt[1])/2.0 + cntrPt[1]
        createPlot.ax1.text(xMid+0.02, yMid, txtString, va="center", ha="center", rotation=0)
    def plotTree(myTree, parentPt, nodeTxt):
        numLeafs = getNumLeafs(myTree)
        depth = getTreeDepth(myTree)
        firstStr = myTree.keys()[0]
        cntrPt = (plotTree.xOff + (1.0 + float(numLeafs))/2.0/plotTree.totalW, plotTree.yOff)
        plotMidText(cntrPt, parentPt, nodeTxt)
        plotNode(firstStr, cntrPt, parentPt, decisionNode)
        secondDict = myTree[firstStr]
        plotTree.yOff = plotTree.yOff - 1.0/plotTree.totalD
```

```
        for key in secondDict.keys():
            if type(secondDict[key]).__name__=='dict':
                plotTree(secondDict[key], cntrPt, str(key))  #recursion
            else:
                plotTree.xOff = plotTree.xOff + 1.0/plotTree.totalW
                plotNode(secondDict[key], (plotTree.xOff, plotTree.yOff),
cntrPt, leafNode)
                plotMidText((plotTree.xOff, plotTree.yOff), cntrPt, str(key))
        plotTree.yOff = plotTree.yOff + 1.0/plotTree.totalD
    def createPlot(inTree):
        axprops = dict(xticks=[], yticks=[])
        createPlot.ax1 = plt.subplot(111, frameon=False, **axprops)
        #createPlot.ax1 = plt.subplot(111)
        plotTree.totalW = float(getNumLeafs(inTree))
        plotTree.totalD = float(getTreeDepth(inTree))
        plotTree.xOff =-0.5/plotTree.totalW;
        plotTree.yOff = 1.0;
        plotTree(inTree, (0.5, 1.0), '')
        plt.show()
    def retrieveTree(i):
        listOfTrees =[{'S': {0: {'NP':{'':'Alice'}}, 1: {'VP': {'':
{'V':{'':'chased'}}, ' ': {'NP':{'':'the', ' ':'rabbit'}}}}}}]
        return listOfTrees[i]
    myTree=retrieveTree(0)
    createPlot(myTree)
```

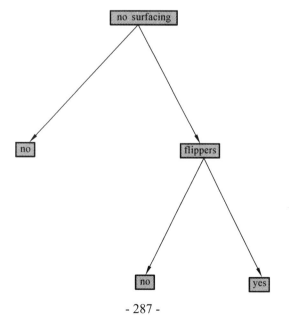

7.4.3 树的遍历

使用递归函数来遍历树是标准的做法。使用一种叫作动态类型的技术检测 t 是一棵树，如定义 t.label()。

【例 7-7】 递归函数遍历树。

```
>>> from nltk import Tree
>>> def traverse(t):
...     try:
...         t.label()
...     except AttributeError:
...         print t,
...     else:
...         #Now we know that t.label() is defined
...         print '(', t.label(),
...         for child in t:
...             traverse(child)
...         print ')',
...
>>> t=Tree('flippers', ['no', 'yes'])
>>> traverse(t)
( flippers no yes )
>>> t.draw()
>>> t
Tree('flippers', ['no', 'yes'])
from nltk import Tree
def traverse(t):
    if not hasattr(t, 'node'):
        print t,
    else:
        print '(', t.node,
        for child in t:
            traverse(child)
        print ')',
t=Tree('flippers', ['no', 'yes'])
traverse(t)
t.draw()
from nltk import Tree
def traverse(t):
```

```
    if not isinstance(t, Tree):
        print t,
    else:
        print '(', t.label(),
        for child in t:
            traverse(child)
        print ')',

t=Tree('flippers', ['no', 'yes'])
traverse(t)
t.draw()
```

7.5 命名实体识别

命名实体是确切的名词短语，指示特定类型的个体，如组织、人、日期等。表 7-3 列出了一些较常用的命名实体类型。

表 7-3 常用命名实体类型

NE 类型	例子
组织	WHO
人	President Obama
地点	Murray River
日期	June，2008-06-29
时间	two fifty am，1:30 p.m.
货币	175 million Canadian Dollars
百分数	twenty pct，18.75%
设施	Washington Monument（纪念碑）

命名实体识别系统的目标是识别所有文字提及的命名实体，可以分解成两个子任务：确定 NE 的边界和确定其类型。命名实体识别经常是信息提取中关系识别的前奏，它也有助于完成其他任务，例如，在问答系统中，试图提高信息检索的精确度，不是返回整个页面而只是包含用户问题答案的那些部分。大多数 QA 系统利用标准信息检索返回的文件，然后尝试分离文档中包含答案的最小文本片段，现在假设问题是 Who was the first President of the US? 被检索的一个文档中包含下面这段话。

```
The Washington Monument is the most prominent structure in Washington,
D.C. and one of the city's early attractions. It was built in honor of
George Washington, who led the country to independence and then became
its first President.
```

分析问题时我们想到答案应该是 X was the first President of the US 的形式，其中 X 不仅是一个名词短语，也是一个 PER 类型的命名实体，应该忽略段落中的第一句话，虽然它包含 Washington 的两个出现，命名实体识别应该告诉我们，它们都不是正确的类型。

我们如何识别命名实体呢？一个办法是查找一个适当的名称列表，例如，识别地点时可以使用地名辞典。然而，盲目这样做会出问题，人或公司名称的情况更加困难，任何这些名称的列表都肯定覆盖不全，每天都有新的公司出现。如果正在努力处理当代文本或博客条目，使用名称辞典查找来识别众多实体是不可能的，困难的另一个原因是许多命名实体措辞有歧义，May 和 North 可能分别是日期和地点类型命名实体的，但也可以都是人名，更大的挑战来自如 Stanford University 这样的多词名称和包含其他名称的名称，因此，在命名实体识别中，我们需要能够识别多标识符序列的开头和结尾。

命名实体识别是一个非常适合用基于分类器类型的方法来处理的任务，这些方法我们在名词短语分块时看到过，特别是，可以建立一个标注器，为使用 IOB 格式的每个块都加了适当类型标签（句子中的每个词加标签）。下面是 CONLL2002 荷兰语训练数据的一部分。

```
Eddy N B-PER
Bonte N I-PER
is V O
woordvoerder N O
van Prep O
diezelfde Pron O
Hogeschool N B-ORG
. Punc O
```

在上面的表示中，每个标识符一行，与它的词性标记及命名实体标记一起。基于这个训练语料，可以构造一个可以用来标注新句子的标注器，使用 nltk.chunk.util.conlltags2tree()函数将标记序列转换成一个块树。

NLTK 提供了一个已经训练好的可以识别命名实体的分类器，可以使用函数 nltk.ne_chunk()访问。如果设置参数 binary=True，那么命名实体只标注为 NE，如（NE U.S./NNP），否则，分类器会添加类型标签，如（GPE U.S./NNP）。

```
>>> from nltk.corpus import treebank
>>> from nltk import ne_chunk
>>> sent=treebank.tagged_sents()[22]
>>> print(ne_chunk(sent, binary=True))
(S
  The/DT
  (NE U.S./NNP)
  is/VBZ
  one/CD
  of/IN
```

```
the/DT
few/JJ
industrialized/VBN
nations/NNS
that/WDT
*T*-7/-NONE-
does/VBZ
n't/RB
have/VB
a/DT
higher/JJR
standard/NN
of/IN
regulation/NN
for/IN
the/DT
smooth/JJ
,/,
needle-like/JJ
fibers/NNS
such/JJ
as/IN
crocidolite/NN
that/WDT
*T*-1/-NONE-
are/VBP
classified/VBN
*-5/-NONE-
as/IN
amphobiles/NNS
,/,
according/VBG
to/TO
(NE Brooke/NNP)
T./NNP
Mossman/NNP
,/,
a/DT
professor/NN
```

```
    of/IN
    pathlogy/NN
    at/IN
    the/DT
    (NE University/NNP)
    of/IN
    (NE Vermont/NNP College/NNP)
    of/IN
    (NE Medicine/NNP)
    ./.)
>>> print(ne_chunk(sent))
(S
  The/DT
  (GPE U.S./NNP)
  is/VBZ
  one/CD
  of/IN
  the/DT
  few/JJ
  industrialized/VBN
  nations/NNS
  that/WDT
  *T*-7/-NONE-
  does/VBZ
  n't/RB
  have/VB
  a/DT
  higher/JJR
  standard/NN
  of/IN
  regulation/NN
  for/IN
  the/DT
  smooth/JJ
  ,/,
  needle-like/JJ
  fibers/NNS
  such/JJ
  as/IN
```

```
    crocidolite/NN
    that/WDT
    *T*-1/-NONE-
    are/VBP
    classified/VBN
    *-5/-NONE-
    as/IN
    amphobiles/NNS
    ,/,
    according/VBG
    to/TO
    (PERSON Brooke/NNP T./NNP Mossman/NNP)
    ,/,
    a/DT
    professor/NN
    of/IN
    pathlogy/NN
    at/IN
    the/DT
    (ORGANIZATION University/NNP)
    of/IN
    (PERSON Vermont/NNP College/NNP)
    of/IN
    (GPE Medicine/NNP)
    ./.)
>>> ne_chunk(sent)
Tree('S', [('The', 'DT'), Tree('GPE', [('U.S.', 'NNP')]), ('is','VBZ'),
('one', 'CD'), ('of', 'IN'), ('the', 'DT'), ('few', 'JJ'), ('industrialized',
'VBN'), ('nations', 'NNS'), ('that', 'WDT'), ('*T*-7', '-NONE-'), ('does',
'VBZ'), ("n't", 'RB'), ('have', 'VB'), ('a', 'DT'), ('higher', 'JJR'),
('standard', 'NN'), ('of', 'IN'), ('regulation', 'NN'), ('for', 'IN'),
('the', 'DT'), ('smooth', 'JJ'), (',', ','), ('needle-like', 'JJ'),
('fibers', 'NNS'), ('such', 'JJ'), ('as', 'IN'), ('crocidolite', 'NN'),
('that', 'WDT'), ('*T*-1', '-NONE-'), ('are', 'VBP'), ('classified',
'VBN'), ('*-5', '-NONE-'), ('as', 'IN'), ('amphobiles', 'NNS'), (',',
','), ('according', 'VBG'), ('to', 'TO'), Tree('PERSON', [('Brooke',
'NNP'), ('T.', 'NNP'), ('Mossman', 'NNP')]), (',', ','), ('a', 'DT'),
('professor', 'NN'), ('of', 'IN'), ('pathlogy', 'NN'), ('at', 'IN'),
('the', 'DT'), Tree('ORGANIZATION', [('University', 'NNP')]), ('of',
```

```
'IN'), Tree('PERSON', [('Vermont', 'NNP'), ('College', 'NNP')]), ('of',
'IN'), Tree('GPE', [('Medicine', 'NNP')]), ('.', '.')])
    >>> ne_chunk(sent).draw()
    >>> ne_chunk(sent, binary=True).draw()
    >>> sent
    [('The', 'DT'), ('U.S.', 'NNP'), ('is', 'VBZ'), ('one', 'CD'), ('of',
'IN'), ('the', 'DT'), ('few', 'JJ'), ('industrialized', 'VBN'), ('nations',
'NNS'), ('that', 'WDT'), ('*T*-7', '-NONE-'), ('does', 'VBZ'), ("n't",
'RB'), ('have', 'VB'), ('a', 'DT'), ('higher', 'JJR'), ('standard',
'NN'), ('of', 'IN'), ('regulation', 'NN'), ('for', 'IN'), ('the', 'DT'),
('smooth', 'JJ'), (',', ','), ('needle-like', 'JJ'), ('fibers', 'NNS'),
('such', 'JJ'), ('as', 'IN'), ('crocidolite', 'NN'), ('that', 'WDT'),
('*T*-1', '-NONE-'), ('are', 'VBP'), ('classified', 'VBN'), ('*-5',
'-NONE-'), ('as', 'IN'), ('amphobiles', 'NNS'), (',', ','),
('according', 'VBG'), ('to', 'TO'), ('Brooke', 'NNP'), ('T.', 'NNP'),
('Mossman', 'NNP'), (',', ','), ('a', 'DT'), ('professor', 'NN'), ('of',
'IN'), ('pathlogy', 'NN'), ('at', 'IN'), ('the', 'DT'), ('University',
'NNP'), ('of', 'IN'), ('Vermont', 'NNP'), ('College', 'NNP'), ('of',
'IN'), ('Medicine', 'NNP'), ('.', '.')]
```

7.6 关系抽取

一旦文本中的命名实体已被识别,便可以提取它们之间存在的关系。如前所述,通常会寻找指定类型的命名实体之间的关系,进行这一任务的方法之一是,首先寻找所有(X, α, Y)形式的三元组,其中 X 和 Y 是指定类型的命名实体,α 表示 X 和 Y 之间关系的字符串,然后可以使用正则表达式从 α 的实体中抽出我们正在查找的关系。下面的例子搜索包含词 in 的字符串,特殊的正则表达式(?!\b.+ing\b)是一个否定预测先行断言,允许我们忽略如 success in supervising the transition of 中的字符串,其中 in 后面跟一个动名词。

```
    Python 3.7.3 (default, Mar 27 2019, 17:13:21) [MSC v.1915 64 bit
(AMD64)] :: Ana
    conda, Inc. on win32
    Type "help", "copyright", "credits" or "license" for more
information.
    >>> from re import compile
    >>> from nltk.corpus import ieer
    >>> from nltk.sem import extract_rels
    >>> from nltk.sem.relextract import rtuple #Python 2.7.11 |Anaconda
2.5.0 (64-bi
```

```
t)
>>> IN=compile(r'.*\bin\b(?!\b.+ing)')
>>> for doc in ieer.parsed_docs('NYT_19980315'):
...     for rel in extract_rels('ORG', 'LOC', doc, corpus='ieer',
pattern=IN):
...         print(rtuple(rel)) #在更深的地方
...
[ORG: 'WHYY'] 'in' [LOC: 'Philadelphia']
[ORG: 'McGlashan &AMP; Sarrail'] 'firm in' [LOC: 'San Mateo']
[ORG: 'Freedom Forum'] 'in' [LOC: 'Arlington']
[ORG: 'Brookings Institution'] ', the research group in' [LOC:
'Washington']
[ORG: 'Idealab'] ', a self-described business incubator based in'
[LOC: 'Los Ang
    eles']
[ORG: 'Open Text'] ', based in' [LOC: 'Waterloo']
[ORG: 'WGBH'] 'in' [LOC: 'Boston']
[ORG: 'Bastille Opera'] 'in' [LOC: 'Paris']
[ORG: 'Omnicom'] 'in' [LOC: 'New York']
[ORG: 'DDB Needham'] 'in' [LOC: 'New York']
[ORG: 'Kaplan Thaler Group'] 'in' [LOC: 'New York']
[ORG: 'BBDO South'] 'in' [LOC: 'Atlanta']
[ORG: 'Georgia-Pacific'] 'in' [LOC: 'Atlanta']

import re
IN=re.compile(r'.*\bin\b(?!\b.+ing)')
s='success in supervising the transition of'
IN.findall(s)
```

搜索关键字 in 执行得相当不错，虽然它的检索结果也会误报，例如，[ORG: House Transportation Committee]，secured the most money in the [LOC: New York]，一种简单的基于字符串的方法排除这样的填充字符串似乎不太可能。

如前文所述，CoNLL2002 命名实体语料库的荷兰语部分不只包含命名实体标注，也包含词性标注。这允许我们设计对这些标记敏感的模式，如下面的例子所示。show_clause()方法以分条形式输出关系，其中二元关系符号作为参数 relsym 的值被指定。

```
import nltk, re
from nltk.corpus import conll2002
vnv=r"""(is/V| #3rd sing present and
    was/V| #past forms of the verb zijn ('be')
```

```
            werd/V| #and also present
            wordt/V #past of worden ('become')
            )
            .* #followed by anything
            van/Prep #followed by van ('of')
        """
        VAN=re.compile(vnv, re.VERBOSE)
        for doc in conll2002.chunked_sents('ned.train'):
            for r in nltk.sem.extract_rels('PER', 'ORG', doc, corpus=
'conll2002', pattern=VAN):
                #print nltk.sem.relextract.show_clause(r, relsym="VAN")
                print nltk.sem.relextract.show_raw_rtuple(r, lcon=True, rcon=True)
        VAN(u"cornet_d'elzius", u'buitenlandse_handel')
        VAN(u'johan_rottiers', u'kardinaal_van_roey_instituut')
        VAN(u'annie_lennox', u'eurythmics')
        defaultdict(<type 'str'>, {'lcon': '', 'filler': u'is/V op/Prep
dit/Pron ogenblik/N kabinetsadviseur/N van/Prep staatssecretaris/N
voor/Prep', 'objsym': u'buitenlandse_handel', 'objclass': u'ORG',
'objtext': u'Buitenlandse/N Handel/N', 'subjsym': u"cornet_d'elzius",
'subjclass': u'PER', 'rcon': '', 'subjtext': u"Cornet/V d'Elzius/N"})
            ...'')[PER: u"Cornet/V d'Elzius/N"] u'is/V op/Prep dit/Pron
ogenblik/N kabinetsadviseur/N van/Prep staatssecretaris/N voor/Prep'
[ORG: u'Buitenlandse/N Handel/N'](''...
            ...'')[PER: u'Johan/N Rottiers/N'] u'is/V informaticaco\xf6rdinator/N
van/Prep het/Art' [ORG: u'Kardinaal/N Van/N Roey/N Instituut/N'](u'in/Prep'...
            ...u'Door/Prep rugproblemen/N van/Prep zangeres/N')[PER: u'Annie/N
Lennox/N'] u'wordt/V het/Art concert/N van/Prep' [ORG:
u'Eurythmics/N'](u'vandaag/Adv in/Prep'...
```

替换最后一行为 print show_raw_rtuple(r，lcon=True，rcon=True)，这将显示实际的词表示两个NE之间关系以及它们左右默认10个词的窗口的上下文。在一本荷兰语词典的帮助下，可以找出为什么结果 VAN('annie_lennox', 'eurythmics')是误报。

小　结

（1）信息提取系统搜索大量非结构化文本，寻找特定类型的实体和关系，并用它们来填充有组织的数据库，这些数据库就可以用来寻找特定问题的答案。

（2）信息提取系统的典型结构以断句开始，然后是分词和词性标注，接下来在产生的数据中搜索特定类型的实体，最后，信息提取系统着眼于文本中提到的相邻的实体，并试图确定这些实体之间是否有指定的关系。

（3）实体识别通常采用分块器，它分割多标识符序列，并用适当的实体类型给它们加标签，常见的实体类型包括公司、人员、地点、日期、时间、货币等。

（4）用基于规则的系统可以构建分块器，例如，NLTK 中提供的 RegexpParser 类，或使用机器学习技术，如本章介绍的 ConsecutiveNPChunker，在这两种情况中，词性标记往往是搜索块时的一个非常重要的特征。

（5）虽然分块器专门用来建立线性的数据结构，其中没有任何两个块允许重叠，但它们可以被串联在一起，建立嵌套结构。

（6）关系抽取可以使用基于规则的系统，它通常查找文本中连结实体和相关词的特定模式，或使用机器学习系统，通常尝试从训练语料自动学习这种模式。

练 习

1. IOB 格式分类标注标识符为 I、O 和 B，三个标签为什么是必要的？如果我们只使用 I 和 O 标记会造成什么问题？

IOB 因为缺少 B-tag 作为实体标注的头部表示，丢失了部分标注信息，导致处理很多任务时效果不佳。

```
[("the", "DT"), ("little", "JJ"), ("yellow", "JJ"), ("dog", "NN")],
[("the", "DT"), ("cat", "NN")]
```

没有缝隙的相邻块如何识别？如双宾语情况, [I_NP, I_NP, I_NP, I_NP], [I_NP, I_NP]。

```
They give me a pen.
```

2. 写一个标记模式匹配包含复数中心名词在内的名词短语，如 many/JJ researchers/NNS, two/CD weeks/NNS, both/DT new/JJ positions/NNS。通过泛化处理单数名词短语的标记模式，尝试进行操作。

```
r"""NP:{<DT|JJ|NN.*>+} #Chunk sequences of DT, JJ, NN"""
```

3. 选择 CoNLL-2000 分块语料库中三种块类型(NP，VP，PP)之一。查看这些数据，并尝试观察组成这种类型的块的 POS 标记序列的任一模式。使用正则表达式分块器 nltk.RegexpParser 开发一个简单的分块器。讨论难以可靠分块的标记序列。

```
import nltk
#sentence='Joey/NN sold/VBD the/DT farmer/NN rice/NN ./.'
sentence='Nick/NN bre/VBD the/DT computer/NN monitor/NN ./.'
sentence=sentence.split()
sentence=[nltk.tag.str2tuple(s) for s in sentence]
grammar=r"""NP:{<DT|JJ|NN.*>+} #Chunk sequences of DT, JJ, NN"""
```

```
cp=nltk.RegexpParser(grammar)
result=cp.parse(sentence)
result.draw()
```

4. 块的早期定义是出现在缝隙之间的材料。开发一个分块器，将完整的句子作为一个单独的块开始，然后其余的工作完全由加缝隙完成。在你自己的应用程序的帮助下，确定哪些标记（或标记序列）最有可能组成缝隙。相对于完全基于块规则的分块器，比较这种方法的性能和易用性。

5. 写一个标记模式，涵盖包含动名词在内的名词短语，如 the/DT receiving/VBG end/NN，assistant/NN managing/VBG editor/NN。将这些模式加入 grammar，每行一个。用自己设计的一些已标注的句子进行测试。

```
import nltk
nouns=[("the", "DT"), ("receiving", "VBG"), ("end", "NN")]
grammar=r"""
    NP:{<DT|PP\$|NN>?<JJ|VBG>*<NN>}
       {<NNP>+}
"""
cp=nltk.RegexpParser(grammar)
print cp.parse(nouns)
cp.parse(nouns).draw()
```

6. 写一个或多个标记模式处理有连接词的名词短语，如 July/NNP and/CC August/NNP，all/DT your/PRP$ managers/NNS and/CC supervisors/NNS，company/NN courts/NNS and/CC adjudicators/NNS。

```
import nltk
from nltk.tag import str2tuple
grammar=r"""
NP:{<DT|NN.*|PRP\$>+}
   {<NP><CC><NP>}
"""
cp=nltk.RegexpParser(grammar)
nouns=['July/NNP and/CC August/NNP',
'all/DT your/PRP$ managers/NNS and/CC supervisors/NNS',
'company/NN courts/NNS and/CC adjudicators/NNS']
nouns=[i.split() for i in nouns]
nouns=[[str2tuple(j) for j in i] for i in nouns]
for i in nouns:
    print cp.parse(i)
    cp.parse(i).draw()
```

7. 用你之前已经开发的分块器执行下列评估任务。(注意:大多数分块语料库包含一些内部的不一致,以至于任何合理的基于规则的方法都将产生错误)

(1) 在来自分块语料库的 100 个句子上评估你的分块器,报告精度、召回率和 F-量度。

```
import nltk
from nltk.corpus import conll2000
sents=conll2000.chunked_sents()[0]
grammar=r"""
NP:{<DT|NN.*|PRP\$>+}
{<NP><CC><NP>}
"""
cp=nltk.RegexpParser(grammar)
print cp.evaluate([sents])
```

(2) 使用 chunkscore.missed() 和 chunkscore.incorrect() 方法识别分块器的错误,并讨论它。

```
import nltk
from nltk.corpus import conll2000
from nltk.chunk import ChunkScore
correct=conll2000.chunked_sents()[0]
grammar=r"""
NP:{<DT|NN.*|PRP\$>+}
{<NP><CC><NP>}
"""
chunk_parser=nltk.RegexpParser(grammar)
tagged_t=conll2000.tagged_sents()[0]
guessed=chunk_parser.parse(tagged_t)
chunkscore=ChunkScore()
chunkscore.score(correct, guessed)
print chunkscore
chunkscore.missed()
chunkscore.incorrect()
```

(3) 与本章的评估部分讨论的基准分块器比较你的分块器的性能。

8. 使用基于正则表达式的块语法 RegexpChunk,为 CoNLL 分块语料库中块类型中的一个开发分块器。使用分块、加缝隙、合并或拆分规则的任意组合。

9. 有时一个词的标注不正确,如 12/CD or/CC so/RB cases/VBZ 的中心名词。替代手工校正标注器的输出,好的分块器使用标注器的错误输出也能运作。查找使用不正确的标记正确为名词短语分块的其他例子。

10. bigram 分块器的准确性得分约为 90%。研究它的错误,并找出它为什么不能获得 100% 的准确率的原因。实验 trigram 分块,你能够再提高性能吗?

11. 在 IOB 块标注上应用 n-gram 和 Brill 标注方法。不是给词分配 POS 标记，在这里我们给 POS 标记分配 IOB 标记。例如，如果标记 DT（限定符）经常出现在一个块的开头，它会被标注为 B(begin)。相对于本章中讲到的正则表达式分块方法，评估这些分块方法的性能。

12. 挑选 CoNLL 分块语料库中三种块类型之一，写一个函数为你选择的类型做以下任务。
（1）列出与此块类型的每个实例一起出现的所有标记序列。
（2）计数每个标记序列的频率，并产生一个按频率减少的顺序排列的列表；每行要包含一个整数（频率）和一个标记序列。
（3）检查高频标记序列。使用这些开发一个更好的分块器。

13. 在评估一节中提到的基准分块器往往会产生比它应该产生的块更大的块。例如：短语[every/DT time/NN] [she/PRP] sees/VBZ [a/DT newspaper/NN]包含两个连续的块，我们的基准分块器不正确地将前两个结合：[every/DT time/NN she/PRP]。写一个程序，找出这些通常出现在一个块的开头的块内部的标记有哪些，然后设计一个或多个规则分裂这些块。将这些与现有的基准分块器组合，重新评估它，看看你是否已经发现了一个改进的基准。

```
import nltk
sent=[('every', 'DT'), ('time', 'NN'), ('she', 'PRP'), ('sees',
'VBZ'), ('a', 'DT'), ('newspaper', 'NN')]
grammar=r"NP:{<[CDJNP].*>+}"
cp=nltk.RegexpParser(grammar)
print cp.parse(sent)
(S (NP every/DT time/NN she/PRP) sees/VBZ (NP a/DT newspaper/NN))
```

14. 开发一个 NP 分块器，转换 POS 标注文本为元组的一个链表，其中每个元组由一个后面跟一个名词短语和介词的动词组成，如[[('the', 'DT'), ('little', 'JJ'), ('cat', 'NN'), ('sat', 'VBD'), ('on', 'IN'), ('the', 'DT'), ('mat', 'NN')]]转换为[('sat', 'on', 'NP')]。

```
import nltk
texts=[[('the', 'DT'), ('little', 'JJ'), ('cat', 'NN'),
('sat', 'VBD'), ('on', 'IN'), ('the', 'DT'), ('mat', 'NN')]]
grammar=r"""
    NP:{<DT>?<JJ>*<NN.*>}
    VP:{<V.*><IN><NP>}
"""
cp=nltk.RegexpParser(grammar)
ts=[cp.parse(text) for text in texts]
x=[]
def traverse(t):
    if t.node=='VP':
        x.append((t[0][0], t[1][0], t[2].node))
    for child in t:
```

```
            if type(child) is nltk.tree.Tree:
                traverse(child)

    for t in ts:
        traverse(t)
```

15. 宾州树库样例包含一部分已标注的《华尔街日报》文本，已经按名词短语分块。格式使用方括号，我们已经在本章遇到它几次了。语料可以使用 for sent in nltk.corpus.treebank_chunk.chunked_sents(fileid)来访问。这些都是平坦的树，正如我们使用 nltk.corpus.conll2000.chunked_sents()得到的一样。

（1）函数 nltk.tree.pprint()和 nltk.chunk.tree2conllstr()可以用来从一棵树创建树库和 IOB 字符串。写函数 chunk2brackets()和 chunk2iob()，以一个单独的块树为它们唯一的参数，返回所需的多行字符串表示。

（2）写命令行转换工具 bracket2iob.py 和 iob2bracket.py，分别读取树库或 CoNLL 格式的一个文件，将它转换为其他格式（从 NLTK 语料库获得一些原始的树库或 CoNLL 数据，保存到一个文件，然后使用 for line in open(filename)从 Python 访问它）。

16. 一个n-gram分块器可以使用除当前词性标记和n-1个前面的块的标记以外其他信息。调查其他的上下文模型，如 n-1 个前面的词性标记或一个前面块标记连同前面和后面的词性标记的组合。

17. 思考一个 n-gram 标注器使用邻近标记的方式，现在观察一个分块器如何重新使用这个序列信息。例如，有两个任务使用名词往往跟在形容词后面的信息，这会出现相同的信息被保存在两个地方的情况，随着规则集规模增长，这会成为一个问题吗？如果是，找到能解决这个问题的方法。

参考文献

[1] LOPER E，KLEIN E，BIRDS. Natural language processing with Python[M]. Sebastopol, CA: O'Reilly，2007.

[2] 周志华. Machine learning[M]. 北京：清华大学出版社，2016.

[3] MANNING C D，RAGHAVAN P，SCHÜTZE H. Introduction to information retrieval[M]. Cambridge： Cambridge University Press，2009.

[4] PEROVSEK M，KRANJC J，ERJAVEC T，et al. Text flows： a visual programming platform for text mining and natural language processing[J]. Science of Computer Programming，2016： 128-152.

[5] 苏金树，张博峰，徐昕. 基于机器学习的文本分类技术研究进展[J]. 软件学报，2006，17（9）：12.

[6] 代六玲，黄河燕，陈肇雄. 中文文本分类中特征抽取方法的比较研究[J]. 中国信息学报，2004，18（1）：1003.

[7] 李荣路. 文本分类及其相关技术研究[D]. 上海：上海复旦大学，2005.

[8] 许阳，刘功申，孟魁. 基于句中词语间关系的文本向量化算法[J]. 信息安全与通信保密，2014（4）：5.

[9] 奉国和. 文本分类性能评价研究[J]. 情报研究，2011，30（8）：5.

[10] 刘旭. 基于 Python 自然语言处理工具包在语料库研究中的应用[J]. 昆明冶金高等专科学校学报，2015，31（5）：6.